ATOMIC DREAMS

REBECCA TUHUS-DUBROW

ATOMIC DREAMS

THE NEW NUCLEAR EVANGELISTS AND THE FIGHT FOR THE FUTURE OF ENERGY

Algonquin Books of Chapel Hill 2025

Algonquin Books of Chapel Hill / Little, Brown and Company
Hachette Book Group
1290 Avenue of the Americas, New York, NY 10104
littlebrown.com

First Edition: April 2025

Algonquin Books of Chapel Hill is an imprint of Little, Brown and Company, a division of Hachette Book Group, Inc. The Algonquin Books of Chapel Hill name and logo are trademarks of Hachette Book Group, Inc.

Portions of this book previously appeared in the *New Yorker* and the *Washington Post Magazine*.

The publisher is not responsible for websites (or their content) that are not owned by the publisher.

The Hachette Speakers Bureau provides a wide range of authors for speaking events. To find out more, go to hachettespeakersbureau.com or email hachettespeakers@hbgusa.com.

Little, Brown and Company books may be purchased in bulk for business, educational, or promotional use. For information, please contact your local bookseller or the Hachette Book Group Special Markets Department at special.markets@hbgusa.com.

Design by Amy Quinn

ISBN 978-1-64375-315-7 (hardcover); 978-1-64375-758-2 (ebook)

Library of Congress Cataloging-in-Publication Data is available.

Printing 1, 2025

LSC-C

Printed in the United States of America

For my daughter

CONTENTS

ATOMIC DREAMS

INTRODUCTION

Those who could mostly stayed indoors. Throughout much of California, the streets, where heat rose visibly from the asphalt, were largely empty. Those who were outside moved with effort, or not at all. An ice cream vendor trudged down a Los Angeles sidewalk, sweating through his shirt, and children who heard his jingle would emerge from their homes momentarily to buy an icy treat. A few miles away, on Skid Row, a mutual aid group handed out bottles of frozen water to people sheltering from the sun in their tents. About 40 miles northwest, a wildfire crackled through the desiccated brush on a hillside. One by one, seven of the firefighters working to contain it had to be hospitalized, sick from the heat.[1]

It was late summer 2022, and most of California was suffering under a "heat dome"—hot, high-pressure air trapped by the atmosphere as if by a lid. The heat would intensify through the Labor Day weekend and into the next week, breaking records throughout the state. In the San Fernando Valley, the Central Valley, and the desert, temperatures climbed above 110 degrees. "If you have AC, use it," the National Weather Service tweeted. "Fans will not be enough." At the same time, on the freeways, drivers saw digital signs, which normally showed messages about distracted driving or upcoming construction, reading, "EXTREME HEAT. SAVE POWER 4–9 PM. STAY COOL." On August 31, Governor Gavin Newsom declared a state of emergency, exempting power plants from certain environmental restrictions to ensure that they could provide enough energy to meet demand.

That day, in Sacramento, the state's capital, temperatures reached 97 degrees, but lawmakers inside the air-conditioned capitol building got a respite from the heat. August 31 also happened to be the last day of the year's legislative session. And for the state senators, the very last bill on the agenda had a direct bearing on the issues raised by the scorching weather: the knotty nexus between a warming climate and energy consumption. The bill's ramifications also touched on a host of other issues, from radiation hazards to marine life to tribal land rights. At stake were more than a thousand jobs, including, potentially, the jobs of the lawmakers themselves. For all that, the practical repercussions of the vote would be dwarfed by its symbolic significance.

The question before them was the fate of California's last nuclear power plant.

The 40 lawmakers had gathered in the majestic senate chamber, an airy space with floor-to-ceiling windows, white Roman Corinthian columns, and plush crimson carpeting. Their desks, made of ornately carved walnut wood, faced the dais, where an American flag hung from a pole and Senator Connie Leyva stood to administer votes. Throughout the chamber, a smattering of mask-covered faces, as well as signs about social distancing, offered reminders of the pandemic that had disrupted the world for the past two-plus years and had only recently begun to abate.[2]

It was the culmination of a productive legislative session in which members had called out their "aye"s and "no"s on hundreds of bills. The issues on the table were pretty much what you'd expect in California circa 2022. There were bills to strengthen abortion protections and support transgender youth. The Freedom to Walk Act was intended to limit enforcement of jaywalking laws; a human composting bill aimed to allow people to turn their bodies into soil after death. There were also multiple measures stepping up the state's already formidable goals to phase out fossil fuels and curb climate change. One of them codified the state's plan to achieve carbon neutrality by 2045; another established a car-free incentive program; a third prohibited oil and gas drilling within 3,200 feet of homes and schools; yet another provided support for efforts to suck carbon dioxide out of the air.

When the floor session started at 10:12 a.m., the mood was celebratory, as the legislature's Democratic supermajority ensured that most bills would pass. At times it was even self-congratulatory, or, at the very least, self-conscious about California's outsized political influence. Not only were the lawmakers setting policy for the nation's most populous state—its 39 million residents far exceeded Texas's 30 million and Florida's 22 million—they also saw themselves as an inspiration for the rest of the world. There were inevitable references to California as the world's fifth-largest economy, to the delegates from other states and countries who routinely visited the state capitol. In rousing remarks on behalf of one ambitious climate bill, Senator Bob Hertzberg, Democrat of Los Angeles, proclaimed, "California has a pretty good track record of knocking the impossible on its tuchus!" Occasionally dampening this spirit were the handful of beleaguered Republicans, who lamented the state's high housing and electricity costs and noted that some Californians were leaving the state for more affordable regions.

As the day wore on, the energy and focus began to fade, and the senators increasingly turned to their phones and their laptops or chatted with one another, even when their colleagues were giving impassioned speeches. At around midnight, Senate President pro Tempore Toni Atkins, a Democrat from San Diego, led a round of thank-yous to administrative and security staff rather than waiting until the session's end, knowing that everyone would be eager to leave at that time. By 1:00 a.m., almost 15 hours into the session, the senators were exhausted. But they still had to vote on one last measure: SB 846.

The thrust of the 31-page bill was this: it would open the door to extending the life of Diablo Canyon Power Plant, the last operational nuclear facility in California. The plant was almost 40 years old, and the operating licenses for the two reactors were scheduled to elapse in the mid-2020s. Originally, Pacific Gas & Electric (PG&E), the utility that owned it, had planned to apply for a license extension for another 20 years. But in 2016,

PG&E reached a deal with environmental groups, as well as unions representing most of the plant's workers, to shutter it in 2025. This agreement included plans to ramp up energy efficiency and renewables construction to replace the electricity from Diablo Canyon. It had received approval from various state agencies and commissions, and it was widely seen as a model—certainly its architects saw it this way—of how to transition to a future that was both nuclear-free and fossil-free, a California powered by wind and sun and water.

A key player in the deal had been Newsom, the former restauranteur with a gravelly voice and blinding smile who had risen through the ranks of California politics to national prominence over the previous 20-odd years. At the time of the 2016 agreement, he was lieutenant governor as well as a member of the State Lands Commission. Now, as governor, in a move some found courageous and others outrageous, he was pushing this new plan—to press PG&E to apply to renew Diablo Canyon's licenses and continue operations. The bill would provide PG&E with a $1.4 billion loan and expedite its regulatory reviews, exempting it from some environmental reviews altogether. Only months earlier, this reversal had seemed inconceivable. The nuclear age in California had been poised to come to an end. But now the senators were being asked to give it new life.

More than half a century earlier, California was at the center of the burgeoning anti-nuclear movement. In 1964, activists had helped defeat plans for a nuclear plant PG&E was planning to build at Bodega Bay, a hauntingly beautiful harbor 50 miles north of San Francisco. Soon afterward, PG&E proposed a new location at Diablo Canyon, in San Luis Obispo County, one of a diminishing number of undeveloped spots on the Central Coast. The plan led to a decades-long battle with local groups that sprang up to protest the plant's construction, as well as national organizations opposed to nuclear power in general.

Anti-nuclear sentiment surged in the late '60s and the '70s as some segments of the public, often taking cues from dissident scientists, grew worried about the hazards of radioactive waste, potential accidents, and weapons proliferation. Protesters marched, holding up signs reading, "Hell no, we won't glow" and "Better active today than radioactive tomorrow!" The major environmentalist groups Friends of the Earth and Greenpeace were each founded with an anti-nuclear ideology at their core; being an environmentalist became synonymous with being anti-nuclear.

But the objections went beyond environmentalism. "Question Authority," instructed buttons and bumper sticker of the era. Americans were disillusioned by the "best and the brightest" who had led the country into Vietnam; by the crimes of President Nixon; and by the scientific establishment, deeply enmeshed with government and the military. Leaders and experts seemed arrogant, condescending, and, quite often, wrong. No one epitomized this dubious authority more than the atomic establishment, who had reassured the public that nuclear power was safe, that it would be "too cheap to meter,"[3] that accidents were vanishingly unlikely to occur. Similarly, the hulking, heavily guarded plants embodied the concentration of corporate power and the militarization of modern life that troubled so many Americans.

In September of 1981, activists blockaded Diablo Canyon, which was still under construction, sitting down and locking arms in front of the gate outside the facility. Over the course of 14 days, more than 1,900 arrests were made, said to be the highest count ever at an anti-nuclear demonstration. They knew their chances of victory were slim, but, as a spokeswoman for one group told the press, "Most of these people couldn't live with themselves if they didn't try."[4]

The activists ultimately failed to derail the plant's operations—its second reactor came online on March 13, 1986. But in the larger struggle over nuclear power's future, they were more successful. A 1976 state law placed a moratorium on building new reactors in California. Throughout the country, new plant construction began to slow for economic reasons. In the aftermath of the accident at Three Mile Island, Pennsylvania, in March of 1979, dozens of planned nuclear plants around the country were canceled.[5] Community opposition, new safety requirements, cost and schedule overruns, and flat electricity demand were among the factors that dimmed the allure of nuclear for utilities. In the final years of the millennium, the nuclear industry seemed to be sputtering along, its future unclear.

But in the 21st century, a handful of voices began calling for a reconsideration of nuclear energy.[6] It wasn't dirty, they said; nuclear plants do not emit the fine particulate matter or other toxic pollutants associated with fossil fuels. It wasn't dangerous, they said; the risks had been blown far out of proportion. Above all, they stressed, it did not emit greenhouse gases, so it could help address what had emerged as a new existential threat: climate change. They pointed out that as protesters had focused their attention on nuclear reactors, few had objected to the spread of coal-fired plants. To nuclear supporters, this seemed analogous to, say, confiscating all the knives from an enemy, but shrugging as they amassed an arsenal of AR-15s.

These claims were advanced by a growing if somewhat ragtag contingent of gadflies, policy wonks, techno-optimists, engineers, industry employees, wealthy entrepreneurs, Twitter reply guys, and self-identified "ecomodernists" and "climate hawks." They might have been easy to dismiss if they hadn't also included several of the world's most respected scientists.

Together, over time, they amounted to something improbable: a pro-nuclear movement. Not only that—a pro-nuclear movement that was making its case on environmentalist grounds. While far smaller than the anti-nuclear movement had been in its heyday, this new faction began to gain influence in the 2010s. And once again, Diablo Canyon emerged as a focal point. Pro-nuclear advocates rallied to save the plant, in what seemed like a doomed campaign. In both cases, the fight to keep it from opening and the fight to keep it from closing, the plant was more than a single facility; it was a symbol. Activists believed that if they could save Diablo Canyon—the most controversial nuclear plant in anti-nuclear California—it would signal a sea change in the state and in the country, and perhaps even herald a new nuclear age.

I CAME OF AGE IN THE '90S, WHEN NUCLEAR WAS SCARCELY ON THE RADAR. The anti-nuclear movement had receded, presumably because no new plants were getting built—in that sense, it had already triumphed—and the pro-nuclear movement had yet to emerge. My parents were anti-nuclear environmentalists, but nuclear power seldom came up. Rather, what I remember most about my family's environmentalism is an acute awareness of resource use. The convergence of my mother's frugality and her anti-consumerist ethos meant that she bought things almost exclusively secondhand, at garage sales or Goodwill—that is, when she bought things at all. One day I noticed an unfamiliar turquoise towel in the linen closet. "Where did you get this towel?" I asked her. "Oh, I found it in the street," she said.* Our household also had a pervasive reverence for the natural; whether in the case of rice, bread, or paper towels, brown was always better than white.

Elementary school lunchtime was painful. Embarrassed by my crinkled, reused brown paper lunch bags and the whole wheat or pumpernickel bread inside, I coveted the crisp new bags and Wonder Bread all around me. By high school, however, I suppose the years of indoctrination had sunk in. I acquired a reputation for vigilance about wasted paper. To get a rise out of me, a boy in my history class once deliberately ripped a fresh sheet of paper in half; 30 years later, I haven't forgiven that tree murderer.

When, sometime in high school, I first learned about what was then called global warming, I felt a prickling anguish. What bothered me was not only the prospect of catastrophe and chaos, but the more subtle changes, even the seemingly benign ones. The notion that humans could influence the weather, could disrupt the New England seasons that had formed a steady backdrop to my life, meant that the rules I had grown up with no longer applied. (I would later learn that, half a century earlier, some people had a very similar response when they first learned about atomic science.) As the changes started to become noticeable, I became the killjoy who glowered on balmy January days as other people happily broke out their shorts.

By the 2010s, I was under no illusion that we could entirely halt climate change, but I thought: *If we can only stave off the truly civilization-ending calamities, I'll take it.* Now I lived in California, and in 2017, I helped organize an initiative in my neighborhood—a kind of Groupon for solar

* My mom wants me to clarify that she washed it before putting it in the closet.

panels—to make buying rooftop solar easier and more affordable. I assumed that wind and solar were the answer: surveying solar panels on roofs, seeing wind turbines dotting landscapes, gave me a little dopamine hit of hope. But I also started hearing more about a solution to which I had never given much thought, though my impressions of it were vaguely negative. This, of course, was nuclear energy. The reason I started paying attention had to do with the respected scientists mentioned above—or at least one in particular: James Hansen.

In June of 1988, Hansen, a 47-year-old NASA scientist, had testified before Congress during a sweltering heat wave. He warned, "The greenhouse effect has been detected and it is changing our climate now."[7] His testimony received front-page coverage in newspapers across the country, and he is credited with bringing the issue to widespread public attention for the first time. In subsequent years, he had courageously spoken out against government inaction, attending protests and even getting arrested, attracting some criticism from peers who thought scientists ought to remain above the fray. As a 10-year-old, I had been oblivious to his 1988 testimony, but I learned about him later from my parents. In my household, Hansen was a hero.

So I had to pause when I learned that he'd decided to endorse nuclear energy as a means of addressing the climate crisis. In fact, he was a full-throated supporter. In December 2015, he coauthored an opinion piece in the *Guardian* with three other climate scientists: "Nuclear power paves the only viable path forward on climate change," the headline read. They proposed that the construction of 115 reactors per year through 2050 could fully replace fossil fuels and meet global energy needs. "The future of our planet and our descendants depends on basing decisions on facts, and letting go of long-held biases when it comes to nuclear power," they argued.[8]

In the '60s and '70s, the experts had seemed to be leading the world toward nuclear apocalypse and endless war in Vietnam. Scientists in particular, in cahoots with the military-industrial complex, had been responsible for developing the weapons that could annihilate humanity. (Some of them, such as, famously, J. Robert Oppenheimer, had come to nurse doubt about their accomplishments.) But by the 2000s, scientists were the ones who were raising the alarm about climate change; oil companies and Republicans were—it became a rallying cry—"denying the science." Experts, at least in this context, had become the good guys again. If Hansen, as well as several other scientists I had reason to trust, championed nuclear energy, I came to think that I ought to learn more about it. I was also simply curious: Wasn't a pro-nuclear environmentalist an oxymoron? Could it really be true that something that had once threatened to doom us was now needed to save us?

THE LONG FIGHT OVER DIABLO CANYON IS AT THE HEART OF THIS STORY because it has so often been at the center of the larger debate about nuclear power in America. The arguments about this single facility distill, in many respects, the broader conflict. Yet Diablo Canyon is, of course, only one plant, and one that's four decades old. This book is also about some of the people who are designing new types of reactors and working to make the industry more democratic, representative, and responsive to communities, even to redress the harms of the past. Advocates for "nuclear justice" are betting that the fundamental activity of splitting atoms can be disentangled from its historical baggage and reinvented for the 21st century.

More than any other energy source, nuclear fission seems to inspire nearly religious attitudes on both sides of the debate. (As far as I know, there is no other movement that focuses on a single type of electricity generation, not even a "pro-solar" or "anti-coal" movement.) Some pro-nuclear advocates are evangelists who believe that it is all but a panacea. "It's gonna be our solution—to everything," one activist, Heather Hoff, told me. "Like, the only way we can have a utopia is if we do nuclear!" She laughed as she said this, perhaps self-conscious about how it would come across. But she meant it. As I spent time talking with these advocates, I began to see this worldview as an ideology, and I began to think of this ideology as "nuclearism." When I looked it up, I found that this term had already been coined, during the Cold War, to mean "dependence on or faith in nuclear weapons as the means for maintaining national security," according to *Merriam-Webster*. But I think it can be applied now to vigorous support for civilian nuclear power—not weapons—which is more prominent in today's discourse: faith in nuclear energy as the means for addressing climate change, air pollution, energy security, and energy independence.

Since its inception in the 2010s, the pro-nuclear movement has developed into an idiosyncratic subculture. Its most prominent members, loosely defined, now range from a '60s-era guru to a Brazilian fashion model to a Nobel Prize–winning physicist. To some extent, it has its own beliefs and shibboleths: if you spend enough time with the pro-nuclear crowd, you start to hear the same talking points, the same lingo, references to the same studies. Yet it is not monolithic; on the contrary, it is fractious. Not all nuclear advocates are evangelists, or nuclearists. But they all believe that nuclear has a meaningful role to play in our energy system.

To some, it's impossible to conceive of pro-nuclear advocates who are not industry shills. In my reporting, I found that some advocates do have various kinds of connections to industry; others do not. Those who do would argue that since, as they see it, nuclear energy is a force for good, there's nothing sinister about the industry. I have tried to be transparent, pointing out industry connections where I have found them, to let the reader decide how to interpret them.

This book focuses on the pro-nuclear movement, which has been ascendant in recent years and, unlike its anti-nuclear counterpart, has hardly been covered.[9] But it's impossible to understand the newer movement—or the history of nuclear power itself—without knowing something about the anti-nuclear movement and the context that gave rise to it. So that, too, is part of the story.

If Diablo Canyon is a symbol of nuclear power, nuclear power itself has long been a symbol. Depending on whom you ask, it represents humanity's genius or our hubris—or both. Nuclear attitudes are intimately linked to our stances on scientific expertise, on modernity, on the proper relationship between humanity and nature. They are bound up with questions of whether humans should be more humble or more audacious; whether we should defer to nature's limits or transcend them in pursuit of what is sometimes called an "abundance agenda."[10] Those who support nuclear tend to believe that energy use is good, and that humanity can harness powerful forces in sophisticated ways to meet its own needs and desires. Those who oppose it tend to believe that no matter how benevolent our intentions, something is bound to go wrong, and with nuclear, the stakes are higher than with anything else. "I think the technology is rad," one nuclear opponent told me. "But the humans dealing with it—we all make mistakes."[11]

The questions raised by nuclear energy are not only technical and scientific; they are political and cultural; they are even psychological and epistemic. I do not pretend to have all—or necessarily any—of the answers. But I have tried to tell the stories of some of the people who hold passionate opinions about this peculiar energy source, and a few of the places where fights over its future are being waged. These fights are ultimately about one of humanity's defining challenges: how to reconcile the use of the energy that powers our lives with the health of the planet that is our home.

CHAPTER ONE

THE LAND

SCOTT LATHROP TOOK HIS FIRST TRIP TO DIABLO CANYON IN THE EARLY 1980s, when the plant was under construction. He was just a few years out of college; he had graduated from the California Polytechnic State University (Cal Poly), San Luis Obispo in 1976 with a degree in industrial technology. In October 1979, he'd been hired by the San Luis Obispo Department of Planning and Building to enforce building codes. PG&E needed permits for trailers they were using as temporary office space during the construction phase, and Scott's boss sent him out to the site to inspect them. He remembers driving out on the access road, which hugged the coastline and cut through a sloped landscape of shrubs and sagebrush. Along the way, he stopped at the security kiosk, where they were expecting him.[1]

It had been a bumpy time for PG&E, with a seemingly endless series of troubles surrounding the plant. Construction had begun in 1968. In 1969, an earthquake fault had been discovered a few miles from the site. This had necessitated seismic upgrades and given opponents, who were already protesting vociferously, new ammunition. In March of 1979, the partial meltdown at Three Mile Island added momentum to the anti-nuclear movement around the country, leading to the huge demonstrations in California and elsewhere. Three months after the accident, about 30,000 people attended a rally and concert at a 50-acre Army airstrip just north of San Luis Obispo. Performers included Jackson Browne, Graham Nash, and Bonnie Raitt. Governor Jerry Brown addressed the crowd, which gave him a two-minute standing ovation, chanting, "No nukes! No nukes!" Brown told them, "I personally intend to pursue every avenue of appeal if the Nuclear Regulatory Commission ignores this community."[2]

It would seem that Scott and his family had good reason to protest, too. Their ancestors had lived on the land where the concrete was being poured, as well as the surrounding lands, which PG&E and its subsidiaries had also acquired.[3] The territory occupied 12,000 acres, spanning 14 miles of shoreline on the Central Coast. It was covered in coastal sage and chaparral, interspersed with creeks that drained into the ocean. To the west were steep, wave-battered rock cliffs. To the east the land sloped into hills and ridges. As the elevation increased, the vegetation shifted to grasslands, live oaks, and bishop pines.

And the place was full of birds. Brown towhees, lark sparrows, hummingbirds and finches flitted about. Red-tailed hawks soared; turkey vultures hovered. Mourning doves let out their melancholy hoots.[4]

The people who had lived there were known as the Northern Chumash.[5] Their houses were thatched domes, and the baskets they wove were tight enough to hold water. They derived most of their sustenance from the sea, gathering clams and abalone, a kind of mollusk, and fishing with hooks crafted out of shells. They also hunted black-tailed deer with bows and arrows, and harvested acorns as well as the seeds of bunch grasses and chia. In 1775, three years after Mission San Luis Obispo was established, Pedro Fages, a Spanish lieutenant, described the region in a report. "It is not to be denied that this land exceeds all the preceding territory in fertility and abundance of things necessary for sustenance," he wrote.[6]

On or near the 12,000-acre plot of land that PG&E would later purchase, at least five small villages had once stood. Their populations were decimated by exposure to diseases the Spaniards brought: tuberculosis, measles, pneumonia. Infant mortality was especially high, and few children survived to adulthood, so that each generation was smaller than the last. Over time, more and more of the people who had lived there joined the Mission, which meant moving to the buildings in the town the Spaniards called San Luis Obispo. There were some holdouts, but gradually, remaining in the villages became less tenable. The Spaniards had herds of sheep and cattle grazing on the same plants whose seeds the Northern Chumash had previously gathered for food. Eventually, the villages were empty.[7]

In 1842, when California was still part of Mexico, the Mexican government gave the territory to a private property owner in a land grant. By the 1960s, it belonged to a cattle company, which then sold it to PG&E. Although Scott didn't know it at the time, his ancestral village was just a few miles south of the plant site, on the surrounding PG&E-owned property. It was called *tsɨkyɨw*, and it was located right on the coast, at the mouth of a creek.[8] The Spanish had borrowed the name, translated as "pecho" ("chest" or "breast" in English), to refer to this entire area: they called it the Pecho Coast.

Scott was born and raised in San Luis Obispo, the youngest of six kids. His family moved frequently from rental to rental. When he was about five years old, the family lived on the outskirts of the city, where his father worked for a farmer who grew alfalfa. Scott remembers riding in tractors and jumping on bales of hay. Later, they lived on the south side of the city, and his father secured work as a park ranger for the county. In retrospect, Scott's childhood seemed to him idyllic. "People really didn't know how great they had it," he told me. "Growing up with Hispanics, Blacks, Asians. All poor, trying to make it good." If a kid was getting out of line, he recalls, nearby adults would not hesitate to intervene. "'Hey, Scotty, stop it.' There was respect of parents."

He had Native American ancestry on both sides of the family, but his mother emphasized this heritage more. Some of her family was from the Juaneño Band of Mission Indians, also called the Acjachemen Nation, in Southern California. She made sure that all of her children were listed on the official roll of tribe members. As a result, when Scott was 16, he, his mother, and each of his siblings received a payment of $668.51—a total of $4,679.57 for the family—as part of a land claims settlement authorized by Congress. His mother divvied up the money as she saw fit, giving Scott $1200 to help him buy a 1968 Chevrolet Malibu.

His father's side of the family, the Northern Chumash side, didn't talk about their lineage much. Scott attributes their reticence to fear. They thought it was more prudent to claim Mexican or Spanish ancestry. These relatives, he said, "probably weren't excited to get on any Native list. They would have looked at it more as a hit list."

But, although Scott didn't realize it until later, his father held on to some of the customs of his ancestors. He would gather abalone, which were abundant when Scott was a child, along the rocky coast. "In certain areas in the period of low tide, wading almost up to your shoulders, you could actually pry abalone off the rocks," Scott recalled. They would also dig for clams at Pismo Beach, located about 20 miles southeast of Diablo Canyon, and Oceano Beach, a few miles farther south. And they'd go deer hunting on a hilly 40-acre plot of land outside the city that had been in the family for generations.

The day Scott visited the plant to inspect the trailers, he parked his county vehicle in the parking lot, went through a search, and donned a hard hat and safety glasses. The two enormous containment domes loomed above him. Unit 1 appeared to be complete; workers were still putting the finishing touches on Unit 2.

Construction workers in hard hats were everywhere. One of them happened to be Scott's eldest brother. Another was his brother-in-law. "It was really strange because I was the young pup," he recalls. "They're about ten years older than I, and here they're out there working, like the worker bees, and I come out as the inspector." When they saw him, they ribbed him. "They're like, 'Get out of here!'"

In 1968 and 1969, an archaeologist named Roberta Greenwood was commissioned by PG&E to excavate six sites that were to be destroyed by the construction. Among the materials she and her team exhumed were 66 burials—some whole skeletons, some in fragments. She found artifacts in these graves: whistles made of bone, pitted stones, shell beads.[9] In 1981, about 80 Native Americans and supporters held a vigil at the plant site to protest the construction on the burial grounds.[10]

Scott was vaguely aware of some of this at the time. He thinks one of his cousins attended the vigil. But he didn't participate in any protests against the nuclear plant. "I was not against and I was not in favor of it," he told me. "I was just dealing with reality." The plant was providing jobs in the community, which was struggling economically—jobs for his relatives. The only other major employers in the area were Cal Poly and agricultural concerns. In any case, nobody asked for his opinion or his blessing. This was decades before a series of state laws would begin to recognize Indigenous groups, mandating that utilities and state agencies consult tribes at certain points in their decision-making processes. It was also decades before he would get deeply involved with his tribe, taking on administrative duties and reading up on Native American history. It was before he learned how most tribes felt about nuclear power, and before he formed his own views.

The site was an odd juxtaposition, this industrial infrastructure in the midst of wild land with ocean views. What Scott remembers most is awe at the scale of the construction project. The domes were made of concrete almost four feet thick, reinforced with six layers of steel rebar and a steel liner. "They were still pouring concrete, and it was amazing to me to see the size of rebar and the amount of rebar that was put into that," he says. The domed houses his ancestors had lived in were about 10 feet high; these domes rose to 215 feet.

Several trailer-offices around the parking lot were already in use, and others they still needed to install. The latter required inspection of the concrete foundation. After he inspected the foundation, one of the workers asked him if he'd give the green light. "Oh, yeah," he said, "no problem, I'll sign off on it." The worker called someone on a radio, then more workers materialized almost immediately.

Scott had no inkling that almost four decades later, he would return to the site numerous times in a different capacity. He did not know how strongly he would come to feel about the land, or how much time he would spend thinking and talking about the nuclear plant. "I was kind of naive," he told me. "I wouldn't have had a clue back then."

CHAPTER TWO

"DON'T SHUT IT DOWN!"

On a sunny Friday in June of 2016, a group of several dozen people emerged from the Bay Area Rapid Transit (BART) station at 16th and Mission Streets in San Francisco. Most of them wore T-shirts—some purple, some mint green—with the words "March for Environmental Hope" emblazoned on the front and "Split Don't Emit!" on the back. They wore sunglasses and baseball caps, backpacks and fanny packs. They carried handmade signs decorated with drawings of trees, flowers, and the symbol for the atom—three interlocking ovals, representing the motion of electrons around the nucleus. Making their way through the plaza outside the station, stepping around cooing pigeons, they marched north down the wide sidewalks of this unglamorous stretch of Mission Street. They passed palm trees, murals, a gas station, a dog day care, then walked under a freeway overpass. Holding their signs aloft, they chanted: "Split don't emit, hey! Split don't emit!"[1]

If you were a not very attentive passerby—just registered the signs with images of nature and the word "nuclear"—you might have supposed it was an anti-nuclear protest. But if you looked more closely, you would have been quickly disabused of that notion. The messages on the signs were unequivocal: "Nuclear Is Green." "Nuclear for Our Future." "Nuclear = Natural."

Double takes were, in a way, the point. The marchers were using the familiar tactics of social movements, including the environmental and anti-nuclear movements, to make the case that those movements had erred catastrophically. With their signs and their shirts and their chants, the pro-nuclear activists were claiming to be the true environmentalists, the true defenders of the planet.

Earlier that day, they had gathered in a parking lot in Oakland to kick off the event. A couple of children scampered around; one wore a March for Environmental Hope T-shirt that came down almost to his feet. The organizer and emcee, a genial young man named Eric Meyer, addressed the group.

"I know not all of us are the best singers, and don't like the sound of our own voices when we're shouting things," he said, "but that's how you're heard, and that's how you make an impact, and that's what we're gonna do." (As it happened, Eric himself was a trained opera singer, but he didn't bring that up.) He laid out the stakes as he saw them: "We're the people, right here, who are on the cusp of whether we make it or break it as a society, as a species," he said. "We're the people. We're the first generation to be able to feel the impacts of climate change, and we're the last generation to be able to do anything about it. And, I gotta say, it makes my heart smile to see everybody here that understands that and shoulders that responsibility, and is ready to make a difference today."

Then he led a series of chants. "Don't shut it down! Don't shut it down!" "Save the plant! Save the plan-et! Save the plant! Save the plan-et!"

Three days earlier, on June 21, PG&E had announced the deal with environmental and labor groups to close the Diablo Canyon Power Plant in 2025, rather than renewing its licenses as originally planned. Some of the marchers were employees at Diablo Canyon, which was located about 250 miles south of where they now stood. Some worked at other nuclear plants and had flown in for the occasion. Others, including Eric Meyer, had no connection to the nuclear industry, but had become convinced that nuclear power was the answer to climate change. Historically, environmentalists had organized countless protests to stop the construction of nuclear plants. They had also held protests to shut plants down. This may have been the first march by self-proclaimed environmentalists aiming to *save* a nuclear plant.[2]

Their first stop was the local Sierra Club office, right around the corner from the parking lot. When they arrived at the generic concrete building, Sarah Spath,[3] an engineer who worked at a plant in Western New York, addressed her fellow protesters, as well as any pedestrians who happened to be walking by. "I'm an environmentalist, but you see, what the world calls an environmentalist would disagree with me," she said. She stood on the sidewalk, looking at her notes through her sunglasses, competing every so often with the cuckoo sound of the crosswalk signal. "Frankly, the science is losing. And why? Fear," she said. "Mistruths and media-propagated misconceptions about the perils of nuclear power have made people equate it with a bomb, make people take a step away every time you say, 'I work at a nuke plant.'"

She was getting at a couple of central tenets of the pro-nuclear worldview. These advocates saw themselves as having science, the facts, on their side, while they saw the other side as irrational, emotional, fear-driven. And they believed nuclear was unfairly stigmatized—which meant that some of them, as nuclear workers, were also stigmatized.

"Anti-nuclear environmental organizations call it, just about every time, 'dirty, dangerous' nuclear power," Spath went on. Although the marchers didn't explicitly frame it this way, their rhetoric drew unmistakable parallels to movements that championed the rights of racial and sexual minorities. They sincerely believed that nuclear power was a victim of bias. Solar and wind were carbon-free darlings, while nuclear was shunned. (One sign at the march seemed to allude pointedly to this belief: It read "Clean Energy Equality," against a rainbow background.)

"I've watched over the past few months as environmental groups have lobbied for and celebrated the closure of several well-maintained, perfectly useful nuclear power plants, using fear as their ally," Spath said. "It's time to redefine environmentalism."

THE NOTION THAT NUCLEAR POWER SHOULD BE PART OF THE SOLUTION TO climate change was not entirely new. Several high-profile environmentalists had "come out" as pro-nuclear (again, phrasing that suggests the stigma associated with it).

One of the earliest was James Lovelock, the British scientist best known as the originator of the Gaia hypothesis in the 1970s—the idea that organisms and the Earth's atmosphere and other systems constantly interact with and influence one another. Lovelock posited that, without any consciousness or intention, these processes maintain conditions hospitable to life—that "after life began it acquired control of the planetary environment," as he and a coauthor put it in a 1974 paper.[4] The substance of this theory, and especially its name, sounded New Age-y, and Lovelock was an eccentric, unaffiliated man of science of a type that had nearly vanished, even by the '70s. All of this endeared him to that era's counterculture and made other scientists skeptical. But in time, as he refined the theory, much of it became broadly accepted, incorporated into the discipline called Earth system science.

In May of 2004, Lovelock published an impassioned plea in the *Independent* that took many of his fans aback. "Opposition to nuclear energy is based on irrational fear fed by Hollywood-style fiction, the Green lobbies and the media," he wrote. "Even if they were right about its dangers, and they are not, its worldwide use as our main source of energy would pose an insignificant threat compared with the dangers of intolerable and lethal heat waves and sea levels rising to drown every coastal city of the world."[5]

Other high-profile champions included climate scientist James Hansen, as we've seen, and Stewart Brand, the '60s icon who founded the legendary *Whole Earth Catalog*. Steven Chu, President Obama's first secretary of energy, a Nobel Prize–winning physicist, was also known as a supporter. Their reasons were similar: Nuclear can provide reliable electricity, without depending on weather conditions such as sunshine and wind and without emitting greenhouse gases. Supporters also tended to believe that the dangers of nuclear had been greatly overstated. So, by 2010, a handful of luminaries with green credentials were touting nuclear. But the fact that you could pretty much name them all off the top of your head showed the extent to which this stance was still the exception to the rule.

The marchers wanted to make clear that they didn't consider nuclear a necessary evil; they weren't reluctantly endorsing it as a bridge to an all-renewables future. They were singing nuclear's praises, effusively and sometimes literally. After Spath's remarks in front of the Sierra Club offices, Eric Meyer led the group, in his deep baritone, in an adaptation of "Battle Hymn of the Republic":

> *"Power, power for the people*
> *Power, power for the people*
> *Power, power for the people*
> *The atom makes us strong."*

If their zeal for nuclear set them apart from most environmentalists, their focus in this march—saving an existing plant—distinguished them somewhat from other nuclear supporters. For years, there had been a small community of nerds who were enthralled by the potential of advanced reactors, held up as safer and smaller and cleaner than the existing plants. People got fixated on certain fuels, especially thorium, or particular coolants, especially molten salt. There were plenty of YouTube videos and TED Talks about the virtues of various promising innovations. That was how Eric Meyer, for instance, had initially gotten sucked in. But this march was about preserving the existing fleet, starting with Diablo Canyon. Nuclear is not just about fantasy designs that don't actually exist yet, the advocates were saying; the current reactors are worth celebrating and defending.

At the time, the United States had about 100 nuclear reactors, which supplied about 20 percent of the nation's electricity and more than half its low-carbon electricity. (There are greenhouse gas emissions involved in all forms of energy generation, if you consider the entire life cycle, from resource extraction to waste disposal. But nuclear, like solar and wind, does not directly emit greenhouse gases while generating electricity, and its life-cycle emissions are comparable to those two sources as well.[6] Accordingly, I refer to nuclear as "low-carbon.") Yet due to a confluence of factors—competition from cheap natural gas; costly upgrades required for safety purposes after the 2011 nuclear accident in Fukushima—a number of reactors were scheduled to close prematurely for economic reasons. Since 2013, four had already shut down, and several more retirements had been announced. And although the costs of renewables had fallen, when nuclear plants shut down, the lost electricity had been largely replaced with fossil fuels. According to the U.S. Energy Information Administration, retired plants in California and Florida had been replaced by natural gas, and after a plant shut down in Wisconsin, coal filled the gap.[7]

When the marchers made their way down Mission Street, their destination was the building where the anti-nuclear environmental group Greenpeace had their local offices. Waiting for them out front, under a green awning, was Michael Shellenberger, a provocateur who had a long history of antagonizing environmentalists. He was a telegenic guy who was not one to avoid attention or arguments; growing up in Colorado, he acted in school plays and then joined the debate team. Until recently, he had co-led the Breakthrough Institute, a think tank he'd cofounded with his erstwhile best friend, Ted Nordhaus. But they'd had a falling-out over personal and strategic differences, in part because Shellenberger wanted to focus on saving threatened nuclear plants. After their split, in late 2015, Shellenberger founded a new nonprofit, Environmental Progress, where he'd recently hired Eric Meyer.[8]

Holding a microphone, sleeves rolled up on the white button-down he had opted to wear instead of a march T-shirt, he said, "I can't tell you guys how happy I am right now. Maybe the happiest moment of my life." Someone in the small crowd let out a whoop. "Look, guys, this is the beginning."

Standing next to him were two other celebrities of the pro-nuclear world: Richard Rhodes, the Pulitzer Prize–winning author of *The Making of the Atomic Bomb*; and Gwyneth Cravens, a novelist and former *New Yorker* editor who had published a book in 2007 titled *Power to Save the World: The Truth about Nuclear Energy*. Rhodes, wearing sunglasses and a straw-colored fedora, gesticulated as he recounted interviewing the president of the company that operated the first nuclear plant in the U.S., in the 1950s; even back then, this man had considered nuclear a green technology, Rhodes said—"which of course it is." Cravens's blond hair blew in the wind as she took the microphone. "In the late nineties a nuclear power plant called Shoreham was being built on Long Island," she said. "I came and rallied against the nuclear plant that they wanted to open, along with a lot of other anti-nuclear people. So that was the last time I was at something like this." She ended with, "I'm honored to be here, and I think you guys are great, and many thanks. And go nuclear!"

"We're going to keep Diablo open," Shellenberger said, jabbing the air with his finger. "We're going to make the case to the California people. They're going to do the right thing."

FOR ALL THEIR ENTHUSIASM, THE CHALLENGES FACED BY THE MARCHERS, in terms of communication and optics, were evident. This was partly a matter of who they were as messengers. The fact that some of them worked at nuclear plants meant that their advocacy entailed obvious self-interest, which could undermine their credibility. And their de facto leader, Shellenberger, had already alienated many people with his attacks on environmentalists and his combative personality—a rift dating back to 2004, when he and Nordhaus had coauthored a notorious essay titled "The Death of Environmentalism." He and Nordhaus had gone on to oppose Obama's cap-and-trade bill; to tout what they called the "shale gas revolution" and what most people called "fracking"; and to embrace nuclear power. While some environmentalists thought they contributed a valuable, pragmatic perspective to the discourse, others found them to be disingenuous contrarians. Inside the pro-nuclear movement, attitudes toward Shellenberger were likewise mixed: some saw him as a charismatic leader, others as a self-promoter who was difficult to work with.[9]

The marchers' characterization of nuclear as the underdog was also going to be a hard sell. The nuclear industry was a kind of younger sibling to the bomb, having grown out of the Manhattan Project. It had been aggressively promoted and massively subsidized by the government, and in its early years, those working in nuclear science had acquired a reputation for secrecy and condescension. And if the industry was struggling now, that seemed, to the average environmentalist, to be because its product was, in fact, dangerous and dirty. The claims of anti-nuclear bias and discrimination, with echoes of the civil rights movement, could sound not only dubious but offensive—especially since virtually all of the marchers appeared to be White.

There were several striking parallels between the nascent pro-nuclear movement and the pro-housing "YIMBY" (yes in my backyard) movement that was emerging around the same time, also in the Bay Area. There wasn't much overlap between the two movements, as each focused single-mindedly on its respective cause, but both were reacting against a form of environmentalism that had predominated for decades—a "small is beautiful" ethos that was skeptical of technology and development. In the case of housing, environmental activists had sought to prevent excessive development that they viewed as environmentally destructive. (They were often joined by "NIMBYs," who just wanted to preserve their own neighborhoods as they were.) The result had been the blockage of nearly all new housing construction in the Bay Area, which in turn led to soaring housing costs as well as unintended environmental consequences: New housing had been pushed out to where it was feasible to build, leading to sprawl, long car commutes, and more development in wildfire-prone areas. The YIMBY movement had now begun to protest this state of affairs and push for dense housing development.

Like the nuclear advocates, the YIMBYs believed that the anti-tech, anti-growth ethos of '60s- and '70s-era environmentalism had outlived its usefulness, if it had ever made sense at all. Both of these new movements prioritized human well-being as well as environmental protection, but they also believed the old-style environmentalism was now undermining its own goals. And when the YIMBYs, led by a young math teacher named Sonja Trauss, started showing up at city hearings in 2014, saying they supported more housing—any housing—it was as unfathomable as when this new breed of environmental advocates began protesting plans to shut reactors down.[10]

Both contingents, at this stage, were also largely populated by a certain type: middle- or upper-middle-class, educated, White, left-leaning but not radical, people who were convinced they had seen the light and that others had not yet caught up with them. In both cases, these characteristics sometimes limited their ability to connect with others who in fact shared many of their values and concerns.

And, crucially, both were in the position of aligning themselves with perceived bad guys: greedy developers and the nefarious nuclear industry. For the nuclear advocates, this position was perhaps even trickier than for the housing advocates. After all, nuclear was seen as downright scary, monstrous, evil. How could you enthusiastically promote the technology that was directly responsible for Three Mile Island, for Fukushima, and the most serious accident of all, the 1986 disaster at Chernobyl? How could you love a technology that was kindred to the bombs that incinerated Hiroshima and Nagasaki?

From the Greenpeace office building, the marchers made their way through the Financial District to 111 Sutter, a 22-story historic building with an arched entryway. Inside were the offices of the Natural Resources Defense Council, one of the groups that had signed on to the deal with PG&E. The marchers gathered in front of the stately entrance, and some, including Shellenberger, sat down, blocking the doors, intending to risk arrest. Standing by were two bemused security guards, one of whom started videotaping on his phone as the marchers sang "Battle Hymn of the Atom" again.

After belting out the song, Meyer told the crowd, "We have a couple people that maybe need no introduction but I'm going to introduce 'em anyway: Kristin and Heather, the Mothers for Nuclear." That jarring combination of words was the name of a group the two women, who both worked at Diablo Canyon, had recently cofounded. Kristin Zaitz had a sleeping toddler strapped to her chest in a carrier; the child's limbs and head hung limply, undisturbed as her mother spoke into the microphone. "These people right here should care about the same things that we care about," she cried. "They should care about our children, they should care about the planet, they should care about all of our futures."

Heather Hoff[11] chimed in. "These people up here, they need to change their minds. They're doing the wrong thing for the people and for the planet. And that's why we're all here. We're gonna start changing minds."

The police never showed up, apparently attending to more pressing matters. There was no confrontation with the NRDC staffers, who simply left through a different exit. It was a modest start, but the marchers weren't discouraged. The final destination was Sacramento, where, in four days, some of them would attend a State Lands Commission meeting on the future of Diablo Canyon. They marched to the Embarcadero BART station, returned to Oakland for dinner, and then drove about an hour north to a campsite, where they pitched their tents by a river.

CHAPTER THREE

THE CLUB

IN THE EARLY 1960S, PG&E WAS NOT IN THE HABIT OF SOLICITING INPUT from others when deciding where to site infrastructure. It's no surprise, then, that the utility neglected to consult the descendants of the Northern Chumash people about the plans to build a nuclear plant at Diablo Canyon. What is surprising is that they did consult another group: the Sierra Club. What is even more surprising is that the Sierra Club said yes.

If, by the 2000s, it seemed almost axiomatic that anyone who cared about the environment would oppose nuclear power, that hadn't always been the case. When the civilian nuclear power industry was new, in the late '50s and early '60s, some Sierra Club members were receptive to it. Back then, though, they weren't called "environmentalists," a word that had not yet been coined; they were known as "conservationists." In fact, historians have argued that the debate about nuclear power would help catalyze the transformation of the old-school conservation movement (with its focus on preserving wilderness) into the modern environmental movement (with its broader attention to pollution and public health). The Sierra Club's internal dispute over Diablo Canyon, in particular, proved a pivotal reckoning.

It's hard to overstate the Sierra Club's influence on the environmental movement—it's the oldest and most storied green organization in the country, and remains powerful today. It was founded in 1892 by John Muir, a Scottish-born naturalist and mountaineer, along with several other nature lovers in the Bay Area. Muir rose to fame as a result of his euphoric nature writing, especially about Yosemite Valley, part of the Sierra Nevada, the 400-mile mountain chain whose snowy peaks and gorgeous valleys undulate from north to south over much of the state's interior. "These blessed mountains are so compactly filled with God's beauty, no petty personal hope or experience has room to be," he wrote.[1] By the '60s, the Sierra Club was the country's most prominent conservation group, and the face of the Sierra Club was David Brower. With his bad teeth and shock of white hair, Brower was known for his monomaniacal dedication to saving wilderness. He was also devoted specifically to the Sierra Club, which he had joined as a young hiker and rock climber in 1933. His own wedding ceremony was delayed because a Sierra Club meeting ran late; his wedding gift to his bride, Anne, was a Sierra Club membership.[2]

Most Sierra Club members possessed what one of them called an "aberrant gene"[3] that made them want to spend every spare moment scaling mountains and dangling from sheer cliffs. When Brower joined, the club was still a California-based organization with a few thousand members, focused chiefly on organizing outings and teaching rock-climbing skills. But in the postwar era, they were forced to confront changes afoot in the country. Between 1933 and 1953, the population of the United States grew from about 125 million to 160 million; the population of California doubled, from about 6 million to 12 million. New suburbs were multiplying all over: in 1950, builders started construction on 1.7 million homes throughout the country.[4] And Americans were snapping up the cars, refrigerators, washing machines, vacuum cleaners, blenders, and toasters that defined modern prosperity. This population growth and economic development meant more and more plans that threatened the wilderness that club members treasured—plans for logging, for highways, for dams that would meet water needs and generate electricity. And if there's one thing the Sierra Club hated, it was dams.

From a 21st-century perspective, dams may seem relatively benign. After all, they are a long-standing, low-tech feature of human societies and even occur in nature. If beavers build them, how pernicious can they be? Today, hydroelectric power is touted by many environmentalists as one of the optimal forms of energy: renewable, reliable, and low-carbon.[5] But back then, conservationists saw them as one of the biggest menaces to their cause. Filling canyons and other scenic areas with water—drowning wildlife, submerging unique natural landscapes—was anathema. The last battle of John Muir's life was an unsuccessful campaign to cancel the construction of a dam in Hetch Hetchy Valley in Yosemite National Park. The defeat stung, and it was one that Sierra Club members would remember for decades.

In 1952, to shepherd the organization into a more politically active phase, Brower was hired as the first paid executive director. With single-minded resolve, he took on campaigns to kill dam projects in Dinosaur National Monument and, in the early '60s, the Grand Canyon. He gave speeches, testified in Congress, and published books of stunning nature photography, often by Ansel Adams, another club member. He placed ads in the *New York Times*. "Now Only You Can Save Grand Canyon From Being Flooded . . . For Profit," one proclaimed, and urged readers to contact government officials. One official described the response this way: "I never saw anything like it. Letters were arriving in dump trucks. Ninety-five percent of them said we'd better keep our mitts off the Grand Canyon and a lot of them quoted the Sierra Club ads."[6]

Yet at this point in his life, Brower also believed in compromise. He believed that if you were going to demand conservation of one extraordinary spot, you should propose an alternative where development was acceptable; and if you were going to oppose one form of energy generation, you should endorse a different one. So, for example, as an alternative to the dams he was so committed to halting, he floated the idea of another option: nuclear power.

In 1938, while Brower was busy scaling cliffs and leading backcountry camping trips, European scientists were pursuing new frontiers in the study of radioactivity and the structure of the atom. In December of that year, the German chemist Otto Hahn was puzzling over the results of a lab experiment. When he'd bombarded a uranium atom with a neutron, new elements had formed and energy was released. His longtime collaborator, Austrian physicist Lise Meitner, figured out what had happened: the nucleus of the uranium atom had actually split in half, much like a drop of liquid. Borrowing a term from biology, they named this phenomenon "fission."

From the beginning, this was understood as not just another discovery, but world-changing knowledge with the potential for both benefit and destruction. Word spread rapidly among the global scientific community, many of whose members had made their way to the United States as refugees from wartime Europe. (Meitner, who was Jewish, was in exile in Sweden.) The scientists quickly grasped the implications. If a neutron could split a uranium nucleus in half, liberating the energy within, there was the possibility of initiating a chain reaction. That is, the neutrons let loose from the first split atom would go on to penetrate other atoms, and so on. The resulting release of energy could be immense.

As Albert Einstein put it in a famous letter to President Roosevelt in 1939, "the element uranium may be turned into a new and important source of energy in the immediate future." The letter, which was prompted by worries that the Nazis had acquired the same knowledge, warned, "This new phenomenon would also lead to the construction of bombs, and it is conceivable—though much less certain—that extremely powerful bombs of a new type may thus be constructed."[7] The missive prompted its own chain reaction: the establishment of the Manhattan Project, which led to the development of atomic bombs and, seemingly inexorably, to their use, even after it became clear that Germany was not on track to produce its own.

After the war, scientists and government officials turned their attention to the less dramatic part of Einstein's letter: the "new and important source of energy." There was a widespread eagerness to take advantage of a new technology that held such enormous potential; there was also a feeling of almost desperation for the productive use of fission to redeem the initial devastation. In a 1946 newspaper article, Meitner was quoted as saying, "We must not be led into drawing too pessimistic conclusions just because the first use to which atomic energy was put happened to be in an engine of destruction. We must look at it as a revolutionizing scientific discovery; but perhaps, even so, only the first step on the road to something greater and still more valuable—mastering the art of using atomic energy for the benefit of mankind."[8] This sentiment was echoed in President Eisenhower's famous "Atoms for Peace" speech at the United Nations headquarters in New York in 1953.

At that point, the public was mostly on board with nuclear power. (The terms "atomic" and "nuclear" can be used pretty much interchangeably; the former was more common back then.) Conservationists, in fact, had particular reasons to support it, largely because it promised to offer an alternative to the dams they loathed. It was also considered cleaner than fossil fuels. The risk of climate change was not yet appreciated in the '50s—the first studies on the rising concentration of atmospheric CO_2 would begin in Hawaii in 1958—but air pollution from fossil fuels was an obvious problem. Brower initially had "high hopes," as he later put it, for atomic energy.[9]

In 1957, the Shippingport Atomic Power Station, the first full-scale commercial nuclear power plant, came online in Pennsylvania, not far from Pittsburgh. (Part of the motive there, as Richard Rhodes would point out years later at the pro-nuclear march, was the intense coal pollution in that area.) By 1962, 14 reactors were either operating or under construction.[10] Nuclear plants were expensive to build, but various kinds of government support were available, and the plants were expected to become increasingly economical as more were constructed. Plus, a big part of the appeal was simply that nuclear was the hot (pun unavoidable) new thing. In a TV documentary, Rhodes described it this way: "It became something one utility's president would brag to his pal over golf . . . 'We've got one of those nuclear reactors to make our electricity, Joe—how about you?'"[11]

PG&E, WHOSE ROOTS DATED BACK TO THE MID-1800S, WAS THE RESULT OF mergers among several companies. By the 1960s, it served most of California, and it was eager to get in on the action. Its leaders projected a surge in electricity demand due to California's population growth and increasing energy consumption. In the early '60s, PG&E had already built one small reactor in the Bay Area, primarily a research and training facility, with another small one under construction in Humboldt Bay.[12]

Then the utility announced plans for a larger power plant, with multiple reactors, at Bodega Bay, a remote fishing village 50 miles north of San Francisco. The area was sparsely populated by fisherman and dairy farmers. Alfred Hitchcock, who shot his film *The Birds* there, rhapsodized: "The ocean to the west, the broad expanse of bay to the east and south, the dramatic green hills rising from the water's edge, and the picturesque little bay community nestled against the shore made Bodega Head a vantage point I should like to revisit again and again in the years to come."[13] In an irony that would become common, the source of its charm—its adjacency to the water, its remoteness—also increased its appeal as a site for a nuclear plant.

The Sierra Club board was against the plan—not because of blanket objections to nuclear energy but because it clashed with their mission to preserve "scenic resources." On September 7, 1963, the board approved a "Sierra Club Policy Relative to Nuclear Power Plants": "The Sierra Club is opposed to the construction of power plants along ocean and natural lake shorelines of high recreation or scenic values." It came with this clarification: "Prior to and since the adoption of this resolution, some have quoted the Sierra Club to be in opposition to the construction of nuclear power plants altogether. Such statements are absolutely without foundation. The Sierra Club never has been opposed to the construction of nuclear power plants per se."[14]

A young staff member named David Pesonen led the fight against the Bodega Bay plant, joining forces with local residents. The protesters ultimately succeeded by bringing in seismic experts to testify that, because it was located on the San Andreas Fault, the site was unsuitable. In October 1964, PG&E canceled the project.[15]

Eager to avoid a repeat of this debacle, the utility came to the Sierra Club before finalizing their plans for another nuclear plant, this time at Nipomo Dunes, in San Luis Obispo County. But this site, too, was magnificent, with distinctive, rippling sand patterns formed by the wind. The Sierra Club wanted it set aside as a park. One member of the San Luis Obispo chapter,

Kathy Jackson, had made it her goal in life to protect the dunes, leading walks there on weekends. It was, it turned out, rather hard to find a spot on the California coast that wasn't a "scenic resource."

After further surveying and real estate negotiations, PG&E proposed yet another site: a wild coastal bluff called Diablo Canyon. PG&E representatives took a handful of members, including Jackson, for a visit. The place was certainly not as singular as the dunes: it was a fairly typical stretch of California coastline. Whether for this reason, or because Jackson was blinded by her devotion to the dunes, or because PG&E took them to a section that was relatively unimpressive, Jackson and the others came away thinking it was a reasonable compromise. At a board meeting in May 1966, Jackson described it to the board as a "gash, a slot in a steep coastal hillside . . . a deep canyon bare except for grass . . . with no trees and an intermittent creek."[16]

The board voted to approve a qualified endorsement, affirming that, as long as certain criteria were met, the club "considers Diablo Canyon, San Luis Obispo County, a satisfactory alternative site to the Nipomo Dunes for construction of a PG&E generation facility[.]"[17]

DURING THIS PERIOD, WHILE FAMILIAR THREATS SUCH AS DAMS AND LOGging were proliferating, Brower and the other conservationists were also learning about novel and insidious dangers of a different order.

On April 26, 1953, physicists who were studying radiation in the town of Troy, New York, noticed after a storm that their instruments registered higher radiation measurements than usual in the air. It turned out that nuclear weapons tests in Nevada had carried "fallout"—radioactive debris—across the country, where it came down in the rain. An isotope called strontium 90 was detected in the fallout. This was particularly worrisome because, scientists knew, strontium 90 behaved like calcium and would move with calcium through the food chain; would make its way into vegetables, milk, and, eventually, teeth and bones.[18] When Americans learned the news, they were horrified, and many were soon demanding a ban on atmospheric testing.

One American who was deeply disturbed by these developments was a marine biologist named Rachel Carson. Carson worked for the U.S. Bureau of Fisheries (later the Fish and Wildlife Service) while moonlighting as a writer. She published a trilogy about the sea, including the National Book Award–winning *The Sea Around Us*, which had been serialized in the *New Yorker* in 1951. These volumes were carefully observed, lyrically written works about the rhythms and patterns of nature, sometimes taking the perspectives of other creatures.

But by the late '50s, Carson was reluctantly shifting her focus to the threats nature was facing. In 1958, she wrote privately: "I suppose my thinking began to be affected soon after atomic science was firmly established. Some of the thoughts that came were so unattractive to me that I rejected them completely. . . . It was pleasant to believe, for example, that much of Nature was forever beyond the tampering reach of man: he might level the forests and dam the streams, but the clouds and the rain and the wind were God's."[19] The breakthroughs of atomic science had disabused her of that belief, and she felt compelled to write about what she was witnessing.

Her subject was not fallout, though, but a less widely known hazard: pesticides, particularly DDT—a chemical whose powerful insecticidal properties were discovered in Switzerland shortly before World War II. DDT was then manufactured prodigiously in the U.S. and, during the war, used to kill mosquitoes and lice to prevent the spread of disease among troops and refugees. When it was no longer needed for military purposes, it was made available for civilian use and was quickly applied to a growing list of problems. DDT was used by home gardeners; swamps in Florida were sprayed to control mosquitoes; wallpaper intended for children's bedrooms was treated with DDT, as was paper marketed as lining for closets and drawers.[20] All of this was just in the first couple of years it was available, and its profligate use continued in the following years.

Carson began looking into the matter in the late '50s and serialized her reports in the *New Yorker* in 1962, when they were also published in book form as *Silent Spring*. The book caught the attention of President Kennedy and became a surprise bestseller. Brower claimed to have read the entire book between legs of a flight while waiting in the Salt Lake City airport.[21]

Silent Spring argued that "chemicals are the sinister and little-recognized partners of radiation in changing the very nature of the world—the very nature of its life." Radioactive substances from weapons tests, along with chemicals sprayed on crops and in forests, were "entering into living organisms, passing from one to another in a chain of poisoning and death."[22]

Carson herself died of cancer in 1964, less than two years after the book was published. But *Silent Spring* would be credited with kick-starting the modern environmental movement that would take shape in the '60s and '70s. While the book's indictment of pesticides is well remembered, the pivotal role of nuclear technology in the work's genesis, and in the text itself, is less often appreciated.

Nuclear technology and DDT raised concerns about the wisdom of scientific interventions more broadly, of tampering with nature, and about the trade-offs between the benefits and costs of modern technology. These two tools had been pursued and deployed aggressively in the urgency of wartime. There had been no time for slow, deliberate consideration; there was also little tolerance for questioning the decisions of the government and the military. But after the war, it seemed to critics that the same incautious ethos was, to a large extent, still operational. Now there was more time and opportunity for observers to pose questions. Both radiation and chemical contamination were invisible, their long-term consequences difficult to ascertain. One or the other would have been unsettling, but together, they caused the pervasive sense that contamination was spreading everywhere; that things were just not right.

When the Sierra Club board voted on the resolution to endorse Diablo Canyon as a location for a nuclear plant, Brower was uneasy with how quickly they'd reached the decision. (As executive director, he was not eligible to vote.) One board member, Martin Litton, had been traveling abroad when that meeting took place. When he learned about the vote, he was apoplectic. A photographer for *Sunset* magazine, Litton had visited Diablo Canyon, and he begged to differ with Kathy Jackson's dismissive characterization of it. Its beauty may have been less unusual than that of the dunes, but its shrubs and grasslands, its rocky cliffs and tide pools were, in their own way, just as splendid. Litton said it was "the last remaining part of the California coast where the coyote roams free, the bobcat is seen almost any morning," and abalone were "so thick they grow on top of each other."[23]

The resolution set off what one member called a "civil war" within the club.[24] Along with concerns about building industrial infrastructure at this particular location, some members were starting to have qualms about nuclear power itself. The fallout controversy had stoked worries about the hazards of radiation. Civilian reactors were supposed to be the beneficial counterpart to atomic weapons—their saving grace—but was that really the case? A handful of dissident scientists began to raise the alarm about radioactive waste and the potential for catastrophic accidents. The Sierra Club had initially seen these issues as outside their purview. But some within the club, including Brower, were starting to think it was time to reconsider that purview, in light of not only this issue but other pressing problems, such as pesticide use and pollution, that were increasingly coming into public consciousness.

Aside from the risk of radioactive contamination, there was another reason Brower was coming to doubt nuclear energy. He saw energy consumption as entwined with population and economic growth, all feeding on one another, together encroaching on wild spaces and poisoning natural systems. In 1968, the Sierra Club's publishing partner, Ballantine Books, released *The Population Bomb*, by Stanford professor Paul Ehrlich, which famously (and erroneously) predicted widespread famine and became a bestseller. In the book's first chapter, Ehrlich described a scene he had witnessed on a vacation in Delhi: "The streets seemed alive with people," he began, which didn't sound so bad. Then he went on, with mounting disgust, "People eating, people washing, people sleeping. People visiting, arguing, and screaming. People thrusting their hands through the taxi window, begging. People defecating and urinating.... People, people, people, people."[25] Brower wrote the foreword. One of his favorite quips was, "How dense can people be?"[26]

And Brower viewed economic growth—which ate up resources in the present without regard to the future—as "a sophisticated device for stealing from our children."[27] He took issue with the whole philosophy of projecting electricity demand and then doing whatever it took—whether building dams or power plants—to accommodate it. If PG&E didn't accommodate the growth, maybe it wouldn't happen. From this perspective, the promised ability of nuclear power to provide nearly limitless quantities of energy was far from a boon—it was a threat.

The Sierra Club leaders at this time were an opinionated bunch. They were overwhelmingly White, well educated, mostly male, and professional: lawyers, engineers, photographers, editors, professors. They often disagreed strongly with one another. The hard-liners, such as Martin Litton, did not believe compromise was ever called for. The moderates, such as Ansel Adams, believed that growth was inevitable and that conservationists should be pragmatic in their efforts to accommodate it while preserving the most extraordinary natural areas. Brower, unlike those who mellow out—or even sell out—with age, went in the opposite direction, becoming increasingly militant over time.

In the late '60s, differences over the direction of the club, Brower's imperious leadership style, and nuclear power all came to a head in the heated fight over Diablo Canyon. There were several attempts to rescind the club's endorsement, but they all failed. Eventually, these fissures culminated in Brower's departure.

In May 1969, the club held a board meeting at the Sir Francis Drake Hotel in downtown San Francisco, an elegant building with a marble staircase and vaulted gold-leaf ceilings. About 400 club members were present, overflowing from a ballroom, along with journalists who had been covering the turmoil at the club.[28] Brower stood before them to give his resignation speech. "There is an enormous amount to be done, for the old addiction to growth will grind up our wilderness, our forests, mountains, and streams in a decade," he said.[29] He announced that he would form a new group, which would later be named Friends of the Earth. This new group would be more radical than the Sierra Club, and firmly anti-nuclear. It would focus on issues such as air and water pollution and pesticides, too. And, gradually, the Sierra Club would follow suit, coming to oppose nuclear and eventually withdrawing its endorsement of PG&E's plant. "With Diablo Canyon," wrote environmental historian Susan R. Schrepfer, "the Sierra Club began to move from a traditional, wilderness conservation agenda toward a comprehensive environmental perspective."[30]

When I think of the Sierra Club during this time, and particularly David Brower, I can't help recalling a famous quote by William F. Buckley. In 1955, in the mission statement for his new conservative weekly opinion journal, *National Review*, he wrote that the magazine "stands athwart history, yelling Stop[.]"[31] This was a credo for conservatism, but it was not totally unrelated to the attitudes of conservationists—as the similarity between their names suggests. Environmentalism, in taking on DDT and fallout and pollution, then expanded the category of objectionable changes in the modern world. Of course, the right wing and environmentalists were lamenting entirely different changes—Buckley called out "radical social experimentation" and big government—and the two factions were often totally at odds. Still, this philosophical orientation may help explain why there has often been tension between the green movement and other strands of the left, which typically strive to change the status quo, not preserve it.[32] Above all, Brower and his allies were yelling "Stop."

And they frequently succeeded. They defeated the proposals for dams at Dinosaur National Monument and the Grand Canyon, as well as numerous other plans for development. Yet, despite the controversy that would only grow, construction of Diablo Canyon broke ground in 1968 and continued apace.

CHAPTER FOUR

THE PLANT

IN 2003, LONG AFTER THE PLANT HAD COME ONLINE, A YOUNG WOMAN named Heather Hoff interviewed for a position at Diablo Canyon. She hadn't particularly wanted to work at a nuclear plant. But she needed a job—a real job with benefits and decent pay and opportunities for advancement. After graduating from Cal Poly in 2002, she'd settled in San Luis Obispo, married her college boyfriend, and drifted among eclectic entry-level gigs. A couple of them embodied the pastoral nature of the regional economy: she shoveled grapes during harvest season at a winery, then worked at a small firm assembling rectal thermometers for cows. By late 2003, she had moved to a more generic position, as a sales clerk at an Express clothing store. She was making $7.50 an hour.

Heather grew up in Arizona in an unconventional family that used resources sparingly: they lived in a trailer with a composting toilet. When she and her sister were little, their father would sprinkle them with a watering can in lieu of showers. Their trailer was surrounded by unpopulated desert land, where the sisters would clamber on rocks, every so often unearthing ancient pottery shards and dodging rattlesnakes. Later, they moved to a house in a nearby town. Above her bed there, Heather hung a poster with a picture of animals on top of the planet Earth. "They were here first," it read.

She didn't know much about nuclear power, but she knew she was an environmentalist, and she had always assumed that nuclear power and environmentalism were at odds. Her parents had also passed on a vague fear of the technology: her mother had been pregnant with her in March of 1979, when the accident at Three Mile Island in Pennsylvania had transfixed the nation. Even though they were far away in Arizona, her parents told her they had been unnerved. After moving to San Luis Obispo, Heather signed up to be on the mailing list of Mothers for Peace, a local anti-nuclear group. They had fought to keep Diablo Canyon from opening, and when that failed, had shifted into a watchdog role, hoping to get it shut down.

On the other hand, jobs at Diablo Canyon were stable and well paying. The plant was one of the county's largest employers, and they were hiring. Heather's undergraduate degree was in materials engineering, so working at a power plant seemed like it could be a good fit. She thought, if nothing else, she might act as a spy, become the Erin Brockovich of nuclear energy. (The corporate villain that Brockovich held to account in the mid-1990s was, in fact, PG&E, for groundwater contamination from a natural-gas compressor station—one of a growing list of scandals that would taint the utility's reputation over the years.)

Swallowing her reservations, Heather applied for a job. In the interviews, she was asked all sorts of intimidating questions. *Can you carry 50 pounds on your back? Can you wear a respirator? Can you work nights and weekends?* She wasn't always sure of the answer, but she said yes every time. Eventually she was hired as a plant operator. As part of her onboarding process, she had to go out to the plant early in the morning for a medical exam. At the security booth guarding the restricted access road that leads to the plant, she stopped for a temporary car pass, then crested a hill in the predawn light. In the distance, she saw the orange glow of the facility's nighttime sodium lights. *Oh my gosh*, she thought. *This is like an alien city out here.*

It was, in fact, a little like a city. More than 1,000 employees worked at the plant, and up to 1,000 more during outages, when they carried out refueling and maintenance.[1] There were plant operators, control room operators, nuclear engineers, civil engineers, mechanical engineers, electricians, machinists, technicians, radiation protection experts, clerical staff, and security guards working different shifts at all hours of the day and night. And, if not exactly alien, it did have distinctive mores that took some getting used to. The place even had, to a certain extent, its own language, some of which was similar to military lingo. Instead of referring, for example, to valve A21, workers would call it Alpha 21, to reduce the possibility of miscommunication; B21 was Bravo 21, and so on. They had acronyms within acronyms within acronyms: AMSAC stood for "ATWS Mitigation System Actuation Circuitry." ATWS stood for "anticipated transient without SCRAM." SCRAM meant the sudden shutting down of the reactor. No one was really sure what that one stood for.[2]

Heather's department, operations, consisted of about 150 people, divided into five crews. The job entailed making the rounds at the facility, checking equipment performance—oil flows, temperatures, vibrations—and hunting for signs of malfunction. She was in training for 14 months. Much of the training took place in an on-site classroom, where she learned about the basics of reactor physics, the mechanics of pumps and valves and motors. She also did rotations in the plant. When she and her coworkers found something amiss, they would fix any problems they could and write up those they couldn't.

The workforce at Diablo Canyon, as in the nuclear industry generally, was overwhelmingly male, and older, so Heather stood out. She tried to hide that she was female, always wearing her hair in a low ponytail and never wearing makeup or jewelry. In general, her coworkers treated her like one of the guys, though they might sometimes allude to her gender. There was one "crusty old guy," Heather recalled, who, when they were on rotations together, would say, "I'm not holding doors for you. This isn't a date!"

In her early years at the plant, she remained wary about the radiation exposure. Her mom had told her she didn't think women should be working at a nuclear plant during their reproductive years. Most of the exposure workers received came from something like walking by pipes that contained radioactive material emitting gamma rays. This type of exposure was considered relatively harmless as long as they limited their time in the vicinity.

But every so often, plant operators had to go into a containment dome "at power"—as opposed to during an outage—exposing them to a different kind of radiation, from the neutrons in the fission reaction. Neutrons are more dangerous: they travel at high speed and easily penetrate clothing and skin. And there was the risk of getting contaminated with those radioactive particles and then spreading them. Heather had to wear a special dosimeter to record her exposure, as well as a protective suit.

As Heather learned in training, the industry had changed significantly since the accident at Three Mile Island that had frightened her parents—indeed, directly as a result. A government report on the incident had harsh words for the industry. Pointing to shortcomings in operator training and control room design, as well as failure to learn across the industry from previous incidents, the report concluded, "given all the above deficiencies, we are convinced that an accident like Three Mile Island was eventually inevitable"—and that, but for luck, the consequences could have been far worse.[3]

In the wake of the accident, the Nuclear Regulatory Commission (NRC) instituted new regulations. Also, following a recommendation from the report, the Institute of Nuclear Power Operations was established to facilitate systematic review of plant operations throughout the industry, ensuring that all plants would learn lessons from one another. In the ensuing years, the industry focused on fostering a strong "safety culture." As a result of the same improvements, plants significantly improved their "capacity factor," or the amount of time they were producing power. (If you were frequently shutting the plant down because of safety issues, you couldn't provide reliable power, one of the major theoretical advantages of nuclear.) In 1975, the average capacity factor was only 55.9 percent. By 2005, it was 89.3 percent and would continue to rise.[4] This was significantly higher than all other energy sources, both fossil and renewable.[5]

The irony, then, was that during the time when the industry was in decline, its reputation in tatters, almost no new plants under construction, its performance had dramatically improved.

One of the few other women at the plant was a civil engineer named Kristin Zaitz. She and Heather didn't work together, but they sometimes bumped into each other at local state parks on weekends with their kids—Heather gave birth to a daughter in 2009, and Kristin had two small children. Nuclear plant workers are not generally known for their environmentalism, but Kristin, like Heather, was a tree hugger. She had grown up in the foothills of the Sierra Nevada with hippie parents in a small house where she shared a bedroom with her three siblings. Her dad would take her backpacking in nearby mountains, and they would sleep under the stars. Even at home, she often slept outside. Her hero was John Muir, the Sierra Club founder.

Her main source of information on nuclear power, before starting at Diablo Canyon, was *The Simpsons*, featuring the rat-infested Springfield Nuclear Power Plant, where green goo oozes out of tanks and three-eyed fish swim in nearby waters. Her views had changed somewhat during a college class at Cal Poly, when she had compared the life-cycle impacts of different energy sources and had been impressed by the small footprint of nuclear plants. She was still suspicious, though. She got an internship at the plant in 2001, during college, but planned to use it as a crash course in civil engineering and then leave after a few months, to pursue some future job designing bridges or roads.

Her coworkers were all men. They taught her steel design, concrete design, and seismic analysis. When they talked specifically about the plant, she would mostly tune it out. After a time, though, on Fridays, two of the men would say, "Get a hard hat—let's go walk around." It was an excuse to get out of the office. They would go to the turbine building and discuss the intricacies of the way certain walls were constructed. They'd walk out in the yard, see the transformers sending electricity out to the grid, and listen to the buzzing sound. She slowly started asking questions and learning more. She thought of these men and the others she worked with as her "adopted fathers." Her wariness gradually gave way to curiosity.

After graduating, she decided to take a full-time job at the plant. A typical day started with a morning meeting, where they discussed the plant's status. Kristin would get a list of assignments or issues to look into. Maybe operators needed to access a certain valve, so she would be tasked with designing a platform. She would call the person who had requested it, gather more details, and collaborate with a more senior engineer to determine the optimal design. Then she and the other engineers would spend the day around the whiteboard. *What shape of steel should we use? Would it be better to use angle iron or a channel section here?*

Sometimes the work was more adventurous. Every so often, she'd inspect one of the containment domes, wearing a white Tyvek suit, climbing a series of ladders to a small platform at the top. Then, donning a harness, she'd rappel down the side of the dome. From there, she had a panoramic view of the lands surrounding the plant—12,000 acres of rugged, nearly pristine coastal terrain—as well as the ocean. To monitor marine life, she periodically went diving in the cove adjacent to the plant, counting fish and recording the numbers on a waterproof clipboard with a grease pen. She would also examine the intake area where the plant sucked in seawater to see if it was occluded by kelp.

For anti-nuclear protesters, the plant's effect on sea creatures was a big concern. As part of its cooling system, the plant sucked in and spit out a staggering 2.5 billion gallons of ocean water per day. When the water was expelled, it was 20 degrees warmer, a "thermal plume" that spread throughout the cove. (It was, however, not radioactive.) Scientists were interested in seeing how this would affect marine life, in part because it might offer insights into how climate-related warming would play out. The assumption was that some species in the cove would migrate north to cooler waters, while other species would come up from the warmer southern waters. It turned out that the changes, while dramatic, were more complex, resulting from shifts in light and food webs. A 2004 study found large declines in kelp and algae, and significant increases in the number of sea snails, sea urchins, and certain fish, such as California sheephead and bat rays.[6] So there was no doubt that the plant's operations were disrupting the ecosystem. To Kristin, though, the sea creatures that were there seemed to be thriving. None appeared to have three eyes.

She loved the work, but being one of the only women did pose challenges. When she returned to work after giving birth to her first child, she realized that the plant was not really set up for nursing mothers. There was no designated room for pumping. "They ended up giving me this super-crappy old trailer," she told me, full of dusty stacks of files. She had to put her breast pump through security every time she entered the plant, and had to explain what it was. "Of course you hear the word 'breast' and it gets weird."

Kristin ended up spearheading the designation of several rooms for pumping, creating a lactation support policy for Diablo Canyon, and joining the local breastfeeding coalition to encourage other local workplaces to follow suit. It was her first taste of the rewards of agitation.

ALTHOUGH HEATHER AND KRISTIN WOULD LATER BECOME THE CLOSEST OF friends and collaborators, they didn't know each other well during the first decade they worked at the plant. But they followed remarkably similar trajectories as their appreciation for nuclear power grew and their fears subsided.

For Kristin, there was a particular "aha" moment. One day, about five years after she'd started working at the plant, she was on the phone with a friend in Washington, D.C., whom she'd met through a professional networking group, North American Young Generation in Nuclear. The friend was telling her about an advocacy effort to support license renewal for nuclear plants in the Northeast. "She told me, 'It's such a big deal because it's the largest source of carbon-free electricity in the Northeast.' I was like, what?" Kristin knew that nuclear was one of several ways to generate climate-friendly energy, but hadn't realized it provided the majority of the nation's total. That call became a flashbulb memory for Kristin: she remembers the cubicle she was sitting in, in a musty outbuilding at the plant, ear pressed to an old gray telephone with a tangled spiral cord.

There was no analogous single moment that allayed her fears; rather, there was a gradual process of both learning about the technology and getting to know and trust her colleagues. The people she was asking questions—health physicists, scientists, engineers—were "so well-intentioned," she said. "After a while, I thought, 'They all can't be lying to me. They live here, they have their families here, and they're not afraid of it, so maybe I should take a cue from that.'"

Heather's evolution was a little more mysterious. When she talks about her years at the plant, it sounds like there were a number of occasions that could have led her to become the Erin Brockovich figure she initially thought she might be.

There was that time she started wondering why so many of her coworkers had gotten cancer. She started making detailed lists of people who had been diagnosed with the disease and who had died of it, and asking others to add names. But when she looked into it more, she realized that 40 percent of all people will have cancer at some point in their lives.[7] (In fact, the literature on the subject is complex. Studies consistently find that nuclear workers actually have a *lower* rate of cancer than average, because of something called the "healthy worker effect"—employees have to meet certain physical criteria to get their jobs. But a couple of large studies of hundreds of thousands of workers, published in the 2010s and 2020s, would find that higher radiation exposure for nuclear workers corresponds to higher cancer rates, suggesting that working at a plant can impose additional risks.[8])

Heather's reaction was similar when, in 2014, she suffered a miscarriage. At first, she suspected that radiation from the plant might be to blame. But then she learned that miscarriages were far more common than she'd realized. For her, researching her suspicions consistently led her to the same conclusion: that while there are risks associated with nuclear energy, there are risks associated with everything. She came to think that "radiation is a lot scarier-sounding than it actually is," she said. "It's a risk balancing. You have to ask, 'Compared to what?'" And she'd come to think the benefits of nuclear—its proven ability to provide large amounts of low-carbon electricity—were worth the risks.

While for most plant workers, the work was just a job—a good job, but basically a way to put food on the table and save for kids' college tuition—for Heather and Kristin, it became something more. They had come to believe that their work was in harmony, not in conflict, with their environmentalist values. They thought nuclear power was making a major contribution to humanity and the planet and had the potential to contribute much more. They had turned from nuclear skeptics into nuclear enthusiasts.

(Paradoxically, this didn't exactly help them fit in better at the plant—most of their coworkers thought their climate-driven passion for nuclear was a little weird.)

In late 2015, though, Heather and Kristin and their colleagues started to pick up on hints of trouble for the future of Diablo Canyon. The current NRC licenses for the two reactors were scheduled to elapse in 2024 and 2025, respectively, and PG&E had begun the process of applying to renew the licenses for another 20-year period. But then managers began to offer "listening and learning sessions" where they talked about "challenges ahead"—such as permits from various state agencies that were also needed.

At first, the women weren't overly concerned. In nuclear, Kristin says, "we're really serious about everything, whether it's a paper cut or a pump that needs to be replaced." These conversations were in keeping with that tone. They didn't start to worry until they learned about a new website called "Save Diablo Canyon." *Save what?* Kristin thought. *That's us—why do we need saving . . . ?*

Heather still received emails from Mothers for Peace, though she now vigorously disagreed with them. Before long, she and Kristin would end up emulating the group, in a sense, while also taking them on as adversaries.

CHAPTER FIVE

"KISS YOUR CHILDREN GOODBYE"

By nature, Jane Swanson wasn't one to make a fuss.[1] She was more comfortable playing the French horn or studying than marching or shouting. A sheltered girl, valedictorian of her high school class, she married her college sweetheart, whom she met at Pomona College, in 1964, and had a baby three months after graduation. When, in 1967, her husband was offered a faculty position in the music department at Cal Poly, they moved to San Luis Obispo and found a small house a few miles from downtown, which was centered around the old Mission church. Jane taught occasional French horn lessons and played in the local orchestra, but mostly stayed home with her daughter. The family had a calm, conventional life, which suited her.

But she kept encountering circumstances that seemed to call for a fuss. The first was the Vietnam War. In 1969, more than half a million American soldiers were stationed in Vietnam, and more than 11,000 would die in that year alone,[2] as would many more Vietnamese soldiers and civilians. That year, a young mother named Joan Stembridge wrote a letter to the editor of the local paper, the *San Luis Obispo County Telegram-Tribune*, saying she wanted to do something to protest the war, but she felt inadequate as an individual. If there were like-minded people, would they please contact her? Then, as one did in 1969, she listed her telephone number. Jane read the letter the morning it was published. She thought, *Wow, this woman is something else.* She was impressed, but she didn't pursue the matter.

Not long afterward, she got calls from several of her friends. *Jane, we went to a meeting, you've got to come to the next one.* She had missed the first meeting of the group that would become Mothers for Peace, but she would seldom miss another. From then on, every few weeks, at least a dozen women met in one of their homes in the evening, sitting on couches and chairs, nibbling baked goods, and plotting how to stop the war. Most of them were young, but Jane, at 25, was among the youngest. With her button nose and bowl haircut, she looked even younger, and lacked the confidence and sophistication she observed in some of the others. Early on, she sat quietly, mainly listening.

The first time she stepped outside her comfort zone was when she went to the local Greyhound bus depot and handed leaflets to the young men who had been drafted and were going to get their physicals. Buses came specifically to take them to the induction center in Los Angeles. The Mothers for Peace would hand them flyers explaining what a conscientious objector was and how to apply for that status. Jane, who had a second child by this point, pushed her double stroller and forced herself to approach these strangers. She was polite and did not say much. Mostly the men looked puzzled but took the flyers and thanked her.

Jane had grown up during the Cold War, when the fear of nuclear war, tangled up with the new fear of weapons testing, suffused pop culture. There was Godzilla, first introduced in a 1954 Japanese film, an ancient sea monster roused back to life by nuclear radiation. The same year brought the black-and-white B movie *Them!*, in which residual radiation from the first atomic bomb test in New Mexico causes common ants to mutate into giant killers. In *Day the World Ended* (1955), survivors of a nuclear war are attacked by a hideous atomic mutant. In *The Incredible Shrinking Man*

(1957), the protagonist is enveloped by a radioactive fog, gradually grows smaller, and ends up living in the only available lodging of appropriate size: a dollhouse. These and other films constituted an entire subgenre about what Susan Sontag called "[r]adiation casualties—ultimately, the conception of the whole world as a casualty of nuclear testing and nuclear warfare[.]"[3]

Jane and her friends had hidden under their desks during drills and lived through the Cuban missile crisis. They had heard about fallout from atmospheric weapons testing, how radioactive debris got into milk and even into babies' teeth. In 1963, President Kennedy had signed the Limited Test Ban Treaty, which outlawed aboveground weapons testing. That was a relief, but tensions with the Soviet Union persisted, and both superpowers were steadily accumulating vast stockpiles of weapons. Once they had banded together, Mothers for Peace, in addition to protesting the Vietnam War, also began working on behalf of nuclear disarmament.

To advance their causes, they wrote letters to the editor, issued press releases, and published educational mailings. They held parties where, in someone's living room, they assembled mailings with their typewritten, mimeographed materials and set up an assembly line across the room for the different tasks: folding the papers, stuffing the envelopes, licking the stamps, writing out the addresses, and applying a rubber stamp with the Mothers for Peace return address. To raise money, they held bake sales and rummage sales in their front yards, placing small notices in the *San Luis Obispo County Telegram-Tribune* to advertise. When they wanted to get the word out about a meeting or another event, they put up posters all over downtown San Luis Obispo, and they called members through a phone tree—each woman had a list of people to call, and those people would also be given a list, and so the branches extended and multiplied.

Plans were underway to build a nuclear power plant 12 miles from Jane's home, but she and her friends were largely oblivious. They barely knew what a nuclear power plant was. One member of Mothers for Peace, however, was married to a Cal Poly physics professor, Gordon Silver, who had concerns about it. In 1970, he told his wife, Sandy, what he knew, and she invited him to talk about it at one of the Mothers for Peace meetings. What he said, as Jane remembers, is that "the radioactive waste of nuclear power plants is exactly the same stuff as radioactive fallout from nuclear weapons atmospheric testing." After some discussion, it wasn't hard to reach a conclusion: "We don't want it falling through the sky, and we don't want it being created here, either."

By 1971, they decided to shift their focus to include protesting the construction of the plant. Mothers for Peace became an official intervenor, as did two of the individual members, Sandy Silver and Liz Apfelberg, which meant they could lodge their objections with the federal Atomic Energy Commission (AEC).

They were not revolutionaries. They were generally sympathetic to the environmentalist and feminist movements of the era, but above all, they were simply, as Jane put it, "anti killing people unnecessarily." Mothers for Peace drew on their moral authority as mothers, and their mandate as such to protect the next generation. "What do you do in case of a nuclear accident," read one of their posters, illustrated by the silhouette of a woman and child embracing. "Kiss your children goodbye."

One of the main reasons radiation, and therefore nuclear power, was so scary—perhaps especially to mothers—was that it was thought to cause genetic damage, which could be passed down to future generations. This belief dated back to the research of geneticist Hermann J. Muller in the 1920s, when he began exposing fruit flies to X-rays. Muller, who would win the Nobel Prize in 1946, also argued that even the most minute exposures to radiation could be harmful.

The AEC, which was responsible for both promoting and regulating civilian reactors, assured the public that nuclear plants would be safe, that radiation exposure would not be an issue. (For much of the '50s, the chairman was Lewis Strauss, famous for orchestrating the downfall of J. Robert Oppenheimer.) But in 1963, the commission asked two scientists, John W. Gofman and Arthur R. Tamplin, to study the potential hazards. Gofman was the director of biomedical research at Lawrence Livermore National Laboratory in the Bay Area, and Tamplin worked with Gofman at the lab. Their research led them to conclude that the dangers of reactors were far graver than the commission had led Americans to believe. When the AEC tried to suppress their findings, the scientists went public with their criticisms, at significant professional risk. As insiders and defectors with no apparent motive other than protecting public health, they enjoyed tremendous credibility.

And they did not go for understatement. At the outset of the 1971 book they coauthored, *Poisoned Power: The Case Against Nuclear Power Plants*, they announced their intention: "It is our purpose to explain how nuclear electricity generation may imperil your health, your life, or your property, without any possibility for redress."[4] Just as the atomic bomb had been different in kind from other weapons, they argued, the risks from nuclear accidents were unique. "We are not speaking of usual 'industrial accidents.' Rather, we are concerned over the hazard of major calamities to human health and life, unparalleled in human history."[5]

There are different kinds of radiation. Visible light, radio waves, and microwaves are forms of radiation, but they are generally benign. The kind of radiation emitted in nuclear plants is known as "ionizing radiation," meaning that it can displace electrons. The fundamental effect of ionizing radiation, they wrote, was "a massive, non-specific *disorganization* or injury of biological cells and tissues."[6] Based on research on rodents, as well as studies of the atomic bomb survivors, they argued that this could lead to cancer. And echoing Muller's theories, they maintained that if the germinal cells were exposed, heritable genetic damage could result.

Their conclusions were maximally alarming. In terms of cancer, "The hazard to this generation of humans from cancer and leukemia as a result of atomic radiation is TWENTY TIMES as great as had been thought previously."[7] In terms of genetic defects, "Changes in the chromosomes of immature sperm or ova cells may be transmitted to all future generations of humans," they warned. "The heredity of man, his greatest treasure, is at stake!"[8] One chapter ends with a radiation hazard sign, preceded by the line: "No amount of ionizing radiation is safe!"[9]

Poisoned Power became something of a bible for anti-nuclear activists, including Jane and her comrades. In October of 1975, Mothers for Peace and 50 local doctors hosted a two-day event, with one keynote speech by Gofman and another by Edward Teller, a Hungarian-born physicist who had participated in the Manhattan Project. Teller had pushed for the creation of the hydrogen bomb, which was far more powerful than the bombs dropped on Hiroshima and Nagasaki and which other scientists, notably Oppenheimer, opposed. Known as the "father of the H-bomb," Teller was relatively sanguine about the risks of radiation. (He even suggested that low doses could be good for you—an argument that would be revived much later.[10]) The local doctors who cosponsored the event wanted to invite the two scientists with opposing views so the public could hear both sides. The event was held at the men's gym at Cal Poly, and an estimated 4,000 people, including Jane, showed up on a Friday night to hear them.

It wasn't that Jane and the other Mothers for Peace didn't trust scientists. It was the scientists—some of them, anyway—who were sounding the alarm.

THROUGHOUT THE NATION, THROUGHOUT THE WORLD, THE ANTI-NUCLEAR movement was growing. In the '70s, there was a wave of direct action—mass arrests during protests at power plant sites, sit-ins on railroad tracks to stop transportation of nuclear materials, and other demonstrations. Luminaries such as Allen Ginsberg, Daniel Ellsberg, and Stokely Carmichael had been spotted at these events. A *Newsday* article on June 12, 1978, noted that "the chant of 'Stop the Nukes' has built into an increasingly militant battle cry across the nation."[11]

The anti-nuclear movement that took shape in the 1970s was a hodgepodge of groups and individuals who saw the issue through different lenses but were united in their disgust. There was no shortage of arguments against nuclear energy, from sophisticated theoretical analyses to viscerally emotional responses. American protesters formed connections with like-minded activists in other countries, from Japan to France to Australia. They didn't buy the distinction between military and civilian uses; a popular date every year for protests at nuclear power plants was August 6, the anniversary of the bombing of Hiroshima.

Although, mercifully, atomic bombs were not used in Vietnam, the distrust sown by that conflict was related to the growing distrust of nuclear energy. In 1968, 50 senior faculty members at MIT signed a statement, which began, "Misuse of scientific and technical knowledge presents a major threat to the existence of mankind." The statement also mentioned disillusionment with the government because of the ongoing war and declared, "We feel that it is no longer possible to remain uninvolved."[12] The Union of Concerned Scientists grew out of this initial statement, and one of the group's major areas of focus was the risk posed by nuclear plants.

A number of celebrities joined the movement. Among them was the pop musician Jackson Browne. "Do you want to risk cancer or birth defects in your kids?" he asked an interviewer. "Do you want radical groups getting hold of uranium? Do you want never to be able to go to L.A. or San Luis Obispo or wherever without some sci-fi suit?"[13] Graham Nash, who performed at anti-nuclear benefit concerts with Browne, captured even more succinctly the raw fear animating the movement. "All that matters to me is securing my future and the future of my child," he told the *Boston Globe* in 1979. "Anybody who wants to bring me down by saying I'm just jumping on a bandwagon can go —— themselves. There won't be any bandwagons if we all fry."[14] In their 1979 song "Power," the Doobie Brothers put it this way: "Just give me the warm power of the sun . . . Just give me the restless power of the wind . . . But won't you take all your atomic poison power away."

The demonstrators came from different segments of society: they included "little old ladies in tennis shoes who were in the 'Ban the Bomb' thing as well as people who were on the fringes of the antiwar movement," according to one activist.[15] The environmental movement was now firmly anti-nuclear, with the Sierra Club, Friends of the Earth, and other new groups, such as Greenpeace, opposing it on the grounds of environmental contamination, as well as "thermal pollution"—the heated water discharged from plants. In 1976, a coalition formed to protest a proposed plant in Seabrook, New Hampshire. In reference to the sea creatures that would be at risk from the plant, they called themselves the Clamshell Alliance. In California, activists followed suit, and a coalition of more than a dozen groups that were protesting Diablo Canyon (including Mothers for Peace) called themselves the Abalone Alliance.

On average, women were more opposed to nuclear power than men. In a 1975 Harris Poll, 70 percent of men expressed support for nuclear energy, compared with 52 percent of women.[16] According to sociologist Dorothy Nelkin, this disparity could not be attributed to lower risk tolerance on the part of women, nor were women more likely to identify as environmentalists. Instead, she suggested in a 1981 paper, "their concern begins with the special effects radiation has on the health of women and on future generations." The female body is particularly vulnerable to radiation, in the form of radiation-induced breast and thyroid cancer, she argued. What was more, the embryo, with its rapid cell division, is highly sensitive to radiation, especially when organs were forming during early pregnancy.[17] "Beyond the specific fears about health risks, however," Nelkin noted, "lies a set of moral and ideological concerns. Women are 'nurturers' or 'caretakers of life,' responsible for opposing life-threatening technologies." This was a generalization, but it certainly reflected how Mothers for Peace saw their mission.

There was also a more radical feminist case. Nuclear power was seen not as the peaceful version of nuclear weapons but as stemming from the same masculine worldview—dominating nature instead of living in harmony with it; unleashing unnatural processes to satisfy exploitative appetites. A representative document of the era, written by activists Susan Koen and Nina Swaim, is called *Aint [sic] No Where We Can Run: Handbook for Women on the Nuclear Mentality*. "Patriarchy is the root of the problem," they wrote, "and the imminent dangers created by the nuclear mentality serve to call our attention to this basic problem of patriarchy."[18] Other female-led groups, whose orientation was more militantly feminist, and less maternal, than Mothers for Peace, included Women Against Nuclear Development (WAND), Spinsters Opposed to Nuclear Genocide (SONG), Dykes Against Nukes Concerned with Energy (DANCE), and Lesbians United in Non-nuclear Action (LUNA).[19]

In short, unless you favored the militaristic, misogynistic, materialistic status quo, you were anti-nuclear. The anti-nuclear movement attracted groups with varying priorities and identities, and it also symbolized broader disaffection with American society. To be anti-nuclear was to believe society had taken a drastically wrong turn—a turn that nuclear power seemed to epitomize.

Despite the protests, construction of nuclear plants continued. By 1979, the U.S. had 72 commercial reactors. That year proved pivotal in the shaping of public opinion toward nuclear power in America.

On March 16, *The China Syndrome*, starring Jane Fonda, Jack Lemmon, and Michael Douglas, was released; the film portrayed safety lapses and a cover-up at a fictional nuclear plant. Twelve days later, one of the two reactors at the Three Mile Island Nuclear Generating Station in southeastern Pennsylvania partially melted down. Confusion reigned for several days as frightened residents heard conflicting statements from the authorities, who were themselves unsure of the accident's severity. On the third day, the state's governor recommended that pregnant women and preschool-aged children living within five miles leave the area. Eventually, an estimated 144,000 people fled.[20]

President Jimmy Carter convened a commission to study the accident, and the commission released its report in October. Although the report identified serious deficiencies in plant operations and the industry as a whole, it concluded that "most of the radiation was contained and the actual release will have a negligible effect on the physical health of individuals." Specifically, "There is a roughly 50 percent chance that there will be no additional cancer deaths . . . and it is practically certain that there will not be as many as five cancer deaths."[21] But the incident highlighted the huge risks involved. Besides, plenty of people simply thought the authorities were lying. It seemed that the Hollywood nightmare, and the predictions of the anti-nuclear movement, were coming true.

By this time, fights over Diablo Canyon had dragged on for more than a decade. But PG&E persisted, and by September 1981, the plant's construction was nearly complete. On September 9, the Nuclear

Regulatory Commission (the successor to the Atomic Energy Commission) approved the utility's security plan for the facility. That milestone provoked the Abalone Alliance to issue a call to supporters to join a blockade of the plant.

On September 12, 1981, protesters began arriving at a campsite in a valley near Diablo Canyon. A village of tents had been erected in a brown field, and a solar-powered loudspeaker blasted music. Members of the alliance were working to set up a medical station in a tepee and to coordinate radio communication. By sunset, about 150 cars were parked in a nearby field, with more steadily streaming in.

On September 15, the blockade started in earnest, with hundreds of protesters sitting down and locking arms on the only paved road to the plant. More than 500 protesters were arrested, in some cases roughly. In the following days, a pattern took hold: arrests in the morning when the workers arrived and then in the afternoon when they departed. On the fourth day, the detainees included Jackson Browne. "I don't want to go to jail but I want to make a statement," he told the press. "My wife's pregnant and we want to have children without genetic defects."

On September 18, more than a dozen protesters were arrested after landing on the beach by rubber raft. By the following day, the action was mostly over, though sheriff's deputies were combing the hills for stragglers who had climbed the fences on the outer perimeter of the site and hidden in the canyons. By September 28, 1,901 arrests had been made.[22]

Meanwhile, Mothers for Peace had continued pursuing any legal avenue they could think of to stop the plant from opening. They held rallies in Mission Plaza in downtown San Luis Obispo; they participated in balloon releases to demonstrate the pattern of radioactive fallout; they organized letter-writing campaigns to elected officials; they attended all of the public meetings and hearings.

In September 1981, just as the blockade was ending, a major design error was discovered. Workers were mistakenly using the same blueprints for seismic safety supports in both units. "The instructions did not explain," a

PG&E spokesman said, "that for the right side was Unit Two, and to get it right for Unit One you had to flip it over."[23] This caused further delays.

Yet PG&E addressed the error, and in April of 1984, Unit 1 started at low power. On August 2, the NRC voted 3–1 to grant PG&E a full-power license for Unit 1. On August 7, Mothers for Peace asked the U.S. Court of Appeals in Washington, D.C., for an injunction on the grounds that the NRC had not taken into account how an earthquake might affect evacuation. To their elated surprise, the Court of Appeals granted a stay on the NRC's full-power decision. But on October 31, the court removed the stay.[24]

After this last-ditch effort failed, Mothers for Peace was out of options to prevent the plant from opening. On May 7, 1985, Unit 1 began commercial operation. Unit 2 began commercial operation the following March, a month before the accident at Chernobyl.

"We didn't win," Jane said many years later, sitting at her dining table. Her hair was now shorter and silver. Her open-plan living and dining room was lined with shelves filled with books on Mozart and Bach, and her husband's double bass was propped up in a corner. "I really thought that we were gonna change history," she said. "Because we were right. We had good lawyers, we had the facts."

I asked if she thought they made a difference at all. "We did make a difference, in many ways. We challenged the Diablo plant and PG&E every time we had a chance, which did slow them down and cost them more money. Mostly they cost themselves money by reading the blueprints wrong. Other utilities looked at that and said, 'I don't think we're gonna do this.'" It's true that anti-nuclear activists were one factor in deterring more plant construction, though not the only one. "I think Mothers for Peace played a role in that, but the biggest role was played by PG&E by screwing it up so badly. I'll give them more credit than I'll give us."

Although the anti-nuclear groups had failed to win the battle to keep Diablo Canyon from opening, they had arguably won the war for public opinion. In the '60s, PG&E had expected that scores of nuclear plants would dot the California coast.[25] At that time, there was very little environmental oversight at the state level. The anti-nuclear movement was instrumental in the birth of what historian Thomas Wellock has called the state's "massive energy bureaucracy," which made the chances of building another plant effectively zero.[26] A 1976 law also outlawed the construction of new plants until the federal government demonstrated that it would follow through on its obligation to deal with the waste. And, as so often, California was a trendsetter. At least a dozen other states, including Massachusetts, New York, Minnesota, and Hawaii, also passed various restrictions and prohibitions on new nuclear plants.[27]

Before I left Jane's house, she wanted to make one last point, a point that is often lost on nuclear proponents who point to the low statistical odds of a disaster. "A lot of risks that we take in life, we get to choose," she said. "We get to choose under what conditions we drive a car, and where we go. We can choose what we eat, we can choose whether we smoke or not. We can make a lot of choices. We can choose if we're going to get on an airplane. They go down once in a while, too. But something as huge as the potential consequences of a bad day at a nuclear plant, that shouldn't be imposed on a big population, or a small population, because a corporation wants to make money. That's exactly what's happening. That's just wrong."

CHAPTER SIX

THE BAD BOYS

IT DIDN'T SEEM LIKE THE KIND OF CONTENT THAT WAS DESTINED TO GO viral. It was a 34-page self-published essay sprinkled with quotes from Heidegger and the *Tao Te Ching*. It was written by an obscure pollster and an equally obscure public relations specialist and posted on a rudimentary new website, breakthrough.org, that no one had ever heard of. And yet, when the authors, Ted Nordhaus and Michael Shellenberger, released it in October 2004, it began to circulate, slowly at first and then rapidly. Debate over it tore through the web "faster than that Paris Hilton home movie," as the environmentally focused outlet *Grist* put it.[1] The authors were inundated with speaking invitations; they were also denounced as arrogant jerks. The fuss had something, though not everything, to do with the title: the authors had decided to call their piece "The Death of Environmentalism."

By this time, the environmental movement had matured into a powerful political force. In the late '60s and early '70s, organizations founded and run by scientists and attorneys—notably the Environmental Defense Fund (EDF) and Natural Resources Defense Council (NRDC)—had sprung up, determined to use legislation and the courts to pursue their objectives. (EDF's motto was "Sue the bastards.") The movement had quickly racked up an impressive series of victories, from a ban on the domestic use of DDT to the creation of the Environmental Protection Agency (EPA) to major legislation such as the 1970 Clean Air Act amendments and the 1972 Clean Water Act. These statutes and others embodied a new kind of environmental law—not setting aside wilderness, as the Sierra Club had traditionally fought for, but regulating pollutants that were contaminating the biosphere and jeopardizing public health. This collection of laws led to clearer skies and healthier rivers throughout the country—what one scholar called an underappreciated "great cleaning."[2]

Although the '80s hardly seemed like an environmental golden age, the movement rose to the occasion to address the era's most salient threats. Whether it was acid rain, caused by sulfur dioxide and nitrogen oxides from power plants, or the hole in the ozone layer, caused by chemicals such as chlorofluorocarbons (CFCs) in aerosols and refrigerants, the environmental community came up with policies to regulate the harmful pollutants and curb the damage.

Then, toward the end of the '80s, yet another environmental problem came to light, dwarfing the rest. In 1988, James Hansen delivered his testimony about the greenhouse effect. The next year, Bill McKibben published *The End of Nature*, the first book about the subject for a general audience. He wrote mournfully of changes that we would begin to witness around us—"and," he predicted, "we will see all too clearly what we have done."[3] In the following years, the issue gradually came to take precedence on the environmental agenda. But the movement was struggling to adapt to a challenge of such unprecedented scope, not to mention one that felt abstract, its gravest consequences in the future. "The Death of Environmentalism," subtitled "Global Warming Politics in a Post-Environmental World," was a salvo in the debate about how best to confront it.

The essay did not mention nuclear power, to which the authors had not yet given much consideration. But Nordhaus and Shellenberger would come to be central figures in the effort to rehabilitate nuclear, to argue that the anti-nuclear movement had profoundly erred. Virtually everyone who would join the pro-nuclear movement in the coming years would have some kind of connection to one or both of them—although some would also clash with one or both of them over differences in values and tone. And in retrospect, the seeds of their pro-nuclear stance could be discerned in the paper that made them semi-famous.

Both authors had connections to the environmental world they were now critiquing, but Nordhaus's ran particularly deep: he came from a kind of environmental aristocracy. His father, Robert Nordhaus, was a government lawyer who had been an author of the 1970 Clean Air Act and other major legislation. His uncle William was a Yale economist who did groundbreaking work on climate change and economic modeling, well before the general public was even aware of the problem. (In 2018, William Nordhaus would win the Nobel Prize in Economic Sciences for this work.[4]) Starting in the '70s, he attempted to assess the "social cost of carbon" and advocated for a tax to account for this "externality," in economics lingo. A tax would impose a cost on emitting greenhouse gases into the atmosphere, which would, the thinking went, gradually reduce fossil fuel consumption while spurring innovation in alternative energy sources. He became one of the leading voices associated with this concept. Bill's work was similar in spirit to the laws drafted by Bob: both sought ways to limit environmental damage associated with economic activity.

Ted grew up on Capitol Hill, a five-minute walk from the House of Representatives. His father often brought colleagues home for dinner, where they would talk shop about energy policy, giving young Ted unusual exposure to the issues. "There weren't a lot of nine-year-olds who knew that, like, Congress was in the middle of trying to establish fuel efficiency standards for automobiles," Ted told me.

As a young adult, Ted started out following more or less in the Nordhaus family footsteps. While taking time off from college at UC Berkeley in the mid-1980s, he worked for the Public Interest Research Group (PIRG), a national network of organizations that had been conceived by Ralph Nader in the '70s. PIRG took up a range of issues, often related to the environmental aspects of consumer rights or public health. For this job, Nordhaus knocked on doors all over the Bay Area, from El Cerrito to San Pablo, asking people to sign petitions and donate money. "It's a shitty, shitty job," he recalled. "It sucked." By his own account, he lacked the charisma that made some of his colleagues naturals at the job, so he had to hone his technique. After residents answered the door, the goal was always to get the clipboard in their hand. "You would hand them the clipboard, then put your hands behind your back."

As he tried to talk to suburbanites about superfunds and consumer utility boards, the overwhelming response was, "I have no idea what the hell you're talking about." Coming from the world of progressive activism, Nordhaus found it eye-opening to learn that most people—at that time, anyway—simply did not think about politics or political issues, especially environmental issues. "They had more of their identity wrapped up in what brand of toilet paper they bought than whether they were environmentalists, or what happened to the toilet paper after they used it," he liked to say. This realization had a lasting influence on his political views.

After returning to Berkeley and graduating, Nordhaus started working as a political operative, and later as a pollster, for various progressive campaigns and organizations. In the mid-1990s, he was hired as executive director of the Headwaters Sanctuary Project, one of several groups that were working to save California's last privately owned large stand of old-growth redwoods, in the far north of the state. That is where he met Shellenberger, who had recently moved to the Bay Area. Shellenberger had started a public relations firm, with assorted left-leaning and environmental groups as clients, and he was working on PR for the Headwaters campaign.

In short order, the two became best friends, spending evenings talking in Nordhaus's backyard in the Berkeley flats. They would drink good wine—Nordhaus, something of an epicure, preferred a local wine merchant called Kermit Lynch—and bounce ideas around. Nordhaus had plenty of other friends, too, but they didn't always get along with Shellenberger, who was "strong medicine," in Nordhaus's words. The two men found that they were intellectually simpatico, with complementary strengths. "It was fun to be around him," Nordhaus told me. "He was just this overflowing fountain of ideas and reactions." Nordhaus played the role of older brother, the "disciplining force." Shellenberger thought Nordhaus was "brilliant," he told me—creative but also a "cold rational thinker." They were both ambitious and eager to make their mark. Shellenberger, in particular, seemed to feel called to greatness. The only question was what form the greatness would take.[5]

So how did they go from fighting to save the redwoods to calling for the death of environmentalism? According to them, they were genuinely frustrated with the big green groups for whom they had each been working in different capacities. And they were heavily influenced by their participation in an initiative called the New Apollo Project.

One of Shellenberger's friends and professional connections was Peter Teague, who had served as an aide to Democratic lawmakers in D.C. and then moved to the Bay Area to work in philanthropy. In 2003, he was director of the environment program at a grantmaking foundation, and he heard from an acquaintance named Bracken Hendricks, a policy wonk who had worked for Al Gore in the White House. Hendricks wanted to pitch Teague a new idea for tackling climate change.

At the time, the most popular idea was to put a price on carbon, which could take the form of either a carbon tax, William Nordhaus's favored approach, or a related idea: cap and trade. The latter would create similar economic incentives in a somewhat more complex way: the government would set emissions limits on the economy as a whole; companies would then buy and sell permits allowing them to emit. A version of this tactic had been successfully used to mitigate acid rain in the '80s.

But the price-on-carbon approach, in Teague's view, was uninspiring. "It was this sort of neoliberal, market-based solution set," Teague told me. "Which all of the foundations had bought into, the environmental community had bought into." He also thought it failed to reflect the magnitude of the problem. "And because the problem was immense and the solution set was small, people tended to think you're either lying about the problem or the solution. They just don't match up."

What Hendricks proposed instead was a large-scale public investment in clean energy. It was essentially industrial policy—that is, significant government funding and oversight to promote the development of a particular economic sector. Industrial policy had often been frowned upon in the U.S., because it was associated with Soviet-style central planning. But Hendricks saw it as a way to unite labor and environmental goals, reviving the moribund American manufacturing base while building the infrastructure to address climate change.[6] When Teague heard the pitch, he thought, *This makes sense. Because we're closing that gap. We're beginning to set a solution that is equal in ambition to the problem.*

He loved the concept but knew that Hendricks would need help on the communications front. Teague immediately thought of Shellenberger. Teague had previously led an LGBTQ community foundation and hired Shellenberger to run a PR campaign. He'd been impressed by what Shellenberger came up with: a newspaper ad campaign targeting Dr. Laura Schlessinger for homophobic comments she'd made on her popular syndicated radio show. After these ads ran, Teague found himself fielding interview requests from media outlets around the world for a month. Teague told Shellenberger about Hendricks's idea, and Shellenberger started developing a proposal with Nordhaus. A couple of labor leaders also joined the effort, as well as Carl Pope, then executive director of the Sierra Club; and Van Jones, the future CNN star, then a young organizer. (His book *The Green Collar Economy* would come out a few years later.) The coalition landed on a vision of a $300 billion investment over 10 years to accelerate the transition to a clean-energy economy and create millions of good jobs. They wanted to sell it as patriotic and entrepreneurial, in the tradition of President Kennedy's moon shot. Hence the name: the New Apollo Project.[7]

Uniting the labor and environmental movements had been a long-standing dream on the left. Together, they promised to be an unstoppable progressive political force. When the leaders of the New Apollo Project reached out to labor, according to Nordhaus, the response was enthusiastic. As for the environmental community, there is some dispute over the reception. Apollo had representation from the Sierra Club on its board, and it received endorsements from multiple other major groups as well. But Nordhaus and Shellenberger perceived the response as lukewarm.

One issue was that according to polls and focus groups that Nordhaus was carrying out, climate change was simply not a priority for most people. So Nordhaus thought they should emphasize what Americans did profess to care about—energy independence, national security, and, especially, jobs—rather than climate. This inclination dated back to his epiphany while knocking on doors. But some environmentalists were understandably wary about downplaying what they saw as the most momentous problem humanity had ever faced.

Another point of contention was that the approach did not entail any guarantee that it would in fact result in reduced greenhouse gas emissions. As opposed to a carbon tax or cap and trade, there were no mechanisms to directly disincentivize emissions. The theory was that building the clean-energy infrastructure would cause fossil fuel consumption to decline, but the proposal did not include binding limits or impose any costs on emissions. In other words, it emphasized carrots rather than sticks.[8]

As Nordhaus and Shellenberger saw it, most of the green groups were stuck in old ways of thinking. They interviewed a couple dozen representatives of those groups—the Sierra Club, Natural Resources Defense Council, Environmental Defense Fund—telling them they were working on a piece about the past 15 years of climate policy. At first they thought they might pitch it to the *New York Times* or the *Atlantic*. But, as they started writing, passing drafts back and forth between them, it grew longer and longer. They decided to publish it themselves, and to call it "The Death of Environmentalism."

In the essay, they acknowledged the past successes of the regulatory approach, the one Nordhaus's father had done so much to make a reality. Pollution control, mandating higher fuel efficiency, cap and trade, and so on had addressed the problems of the previous era: smog, acid rain, ozone-depleting chemicals. But the challenge of global warming, they argued, was different. For one thing, it implicated our entire economy, and was therefore not something that could be solved by tinkering with one facet of the energy system. They quoted Van Jones: "The first wave of environmentalism was framed around conservation and the second around regulation. We believe the third wave will be framed around investment."[9]

Relatedly, Shellenberger and Nordhaus argued that the nature of the challenge called for rethinking what counted as "environmental." "Why, for instance, is a *human*-made phenomenon like global warming—which may kill hundreds of millions of *human beings* over the next century—considered 'environmental'?" they wrote.[10] Channeling Kennedy again, they added, "The arrogance here is that environmentalists ask not what we can do for non-environmental constituencies but what non-environmental constituencies can do for environmentalists."[11]

Perhaps most important, the authors charged that the environmental movement was bogged down in mundane technical solutions, failing to advance an inspiring vision that could mobilize the public—something like, well, the New Apollo Project. They took direct aim at Nordhaus's uncle's life's work: "Who cares if a carbon tax or a sky trust [a similar scheme] or a cap-and-trade system is the most simple and elegant policy mechanism to increase demand for clean energy sources if it's a political loser?" As an alternative to technical policy solutions like a carbon tax or small-bore tweaks like fluorescent lightbulbs, they argued that environmentalists should "tap into the creative worlds of myth-making, even religion . . . to figure out who we are and who we need to be."

In short, Nordhaus and Shellenberger were saying: Focus on the big vision, not on specific policies. Don't scold Americans about what they should value (stopping climate change); meet them where they are and talk about what they do value (jobs). Don't be dour Cassandras; be can-do optimists. ("Imagine how history would have turned out had [Martin Luther] King given an 'I have a nightmare' speech instead.") Also, instead of making dirty energy expensive (through pricing mechanisms like a carbon tax or cap and trade), make clean energy cheap (through massive public investment, similar to the kind that produced the railroads and the internet). They concluded, "We have become convinced that modern environmentalism, with all of its unexamined assumptions, outdated concepts and exhausted strategies, must die so that something new can live."

In October 2004, the pair, in addition to posting the document on their website, produced a printed version. The green cover page featured the Chinese ideogram for "crisis," which, the second page explained, comprises the characters for "danger" and "opportunity." (This was an old chestnut, dating back most famously to a speech, again, by President Kennedy, although it turns out not to be quite accurate in either use.[12]) Nordhaus filled two boxes with hundreds of copies of the printouts, checked them at the airport, and flew to Kauai, Hawaii, where he met up with Shellenberger and Teague, who wrote the essay's foreword. The occasion was the annual meeting of the Environmental Grantmakers Association.

In the open-air lobby at the Hyatt, with breathtaking ocean views, other groups had laid out their annual reports and other literature on tables. Nordhaus and Shellenberger placed their printouts on the same tables, and people started to pick them up as they passed by. Rather predictably, it was not well received. "It landed like a neutron bomb," says Teague. "It was tremendously upsetting to people."

Within the larger environmental community, the reaction was a sense of betrayal—that Nordhaus and Shellenberger had misrepresented them in order to promote themselves. Although the essay was in part a brief for the New Apollo Project, their colleagues there were not pleased, either; the paper threatened to alienate the people they were trying to reach. Soon after the conference, Carl Pope, the Sierra Club executive director who was on the board of Apollo and was also one of the interviewees quoted in the essay, wrote a scathing response, first sent out to conference attendees and later posted on *Grist*. Calling it "unfair, unclear, and divisive," he wrote that, in criticizing environmentalists for lack of creativity and vision, S&N (as he dubbed them) were ignoring voices like Wendell Berry and Terry Tempest Williams. "They interviewed 25 policy people, and then complain that they got only policy expertise from their interviews." He admitted that "we do have some work to do," but (in reference to the essay's title) countered that "dying does not seem a particularly helpful form of that work."[13] Shellenberger and Nordhaus were kicked off the project.

The essay attracted more attention when, the month after its publication, John Kerry lost the 2004 presidential election to George W. Bush. To some, the election results seemed to vindicate the argument that the environmental movement—which had tried desperately to defeat Bush—needed fresh approaches. S&N (as I'll refer to the duo now) had also explicitly applied their argument not only to environmentalism but to the progressive movement more generally, which they argued had devolved into a bunch of siloed "special interests" that were narrowly focused on their own issues and failing to come together in a broad, patriotic coalition.

At least some environmentalists were receptive to the critique. In January 2005, Nordhaus and Shellenberger were invited to attend a climate conference at Middlebury College in Vermont, along with climate activists including Bill McKibben. The event took place on one of those New England days that made it easy to forget about the whole problem: cold, clear blue sky, fresh snow on the ground.[14] At the conference, McKibben did them the favor of dubbing them "the bad boys of American environmentalism." *New York Times* reporter Felicity Barringer was there; her dispatch ran under the headline, "Paper Sets off a Debate on Environmentalism's Future."[15]

The debate was vigorous. Many agreed that the big green groups were out of touch. But S&N were hardly the first to point this out. Others—especially environmental justice advocates—had been saying so for years and had never gotten this kind of attention. S&N had likely generated so much notice in part because they were seen as insiders: White male environmentalists attacking other White male environmentalists (the media loves infighting on the left, *Grist*'s David Roberts noted[16]). Although they may have seen themselves as underdogs—rebelling against the gatekeepers at the well-funded big green groups—others didn't see them that way. And some readers thought they made valid points but in a gratuitously antagonistic manner. In her article, Barringer quoted the executive director of Greenpeace USA, who was at the conference and told her, "These guys laid out some fascinating data, but they put it in this over-the-top language and did it in this in-your-face way."

Indeed, it would be easy to imagine a slightly rejiggered version of the piece with a title like "The Path Forward for the Green Movement" or "The Rebirth of Environmentalism." There were antecedents to some of their arguments in the environmentalist past, and even some of the quotes from the interviewees were in line with their theses. A different set of coauthors might have emphasized this continuity and these points of agreement. But that wasn't S&N's style. As they would go on to demonstrate at the Breakthrough Institute, the think tank they cofounded a couple of years later, they preferred to amplify differences rather than bridge them, and to ignore common ground rather than stressing it. They cast themselves as smart iconoclasts—and they succeeded in capturing attention where a less contrarian take would not have.

That said, there was something in the essay and in their subsequent work that was genuinely different from the dominant strain of environmentalism. As we saw, that movement, dating back to its conservationist roots, had focused on stopping projects (dams, nuclear plants) and limiting damage (pollution, pesticides); in short, restraining human impact. By contrast, Nordhaus and Shellenberger were arguing for a focus on building and creating—that is, unleashing human potential and embracing human aspiration. Whether that shift could effectively address climate change remained to be seen, but it did represent a philosophical departure.

In 2007, Nordhaus and Shellenberger published a book expanding on the essay, titled *Break Through: From the Death of Environmentalism to the Politics of Possibility*, and in November of that year, acclaimed author Michael Pollan hosted a conversation with them on the campus of UC Berkeley. When he'd read "The Death of Environmentalism," he told the audience, "I found their argument very challenging and provocative, as someone who'd been writing about nature for many years and questioning some of the kind of deeper assumptions of the environmental movement in America." The three men sat in chairs on a wooden stage, Shellenberger and Nordhaus to the left of a table with a vase of flame-colored flowers and two plastic water bottles, Pollan to the right. Turnout was good; to accommodate the audience, they held the event in Sibley Auditorium, a large venue in the Bechtel Engineering Center, rather than at the journalism school.[17]

"Death of Environmentalism" had not mentioned nuclear power at all, and *Break Through* did only in passing: the authors wrote that an effective environmental movement would need "to accept and indeed embrace concepts like cleaner coal technologies and perhaps even nuclear energy."[18] By contrast, the book dedicated an entire chapter to defending the offshore wind project proposed for Nantucket Sound. But that evening, when Pollan opened the conversation up to the audience, the first questioner asked for their take on nuclear.

Nordhaus replied: "I'll tell you, we both came up sort of, you know, in the environmental movement and with a sort of strong anti-nuke perspective. And I have to say, you know, just reading the energy science and energy economics literature, it's really hard to find any serious mainstream review of how you get from here to eighty percent reduction in carbon that doesn't include some pretty significant role for nuclear."

He said that since the heyday of the anti-nuclear movement, "the stakes have kind of changed." Now, with the threat of dramatic impacts from climate change, the potential risks of nuclear waste leaking in thousands of years—one of the major worries of the anti-nuclear movement—didn't seem quite as urgent. "I think that's why a lot of folks are at least taking a second look at it."

Soon after that conversation, Nordhaus and Shellenberger became increasingly bullish on nuclear. They were already willing, if not eager, to break with green orthodoxy. Nuclear energy fit seamlessly with their vision of celebrating human capability and ingenuity. Its image—masculine and aggressive—also matched their vibe. They didn't change their minds in a vacuum, however. The evolution of their views was heavily influenced by a handful of scientists and others who, during this same period, had been among the folks taking a second look at it.

CHAPTER SEVEN

CONVERSIONS

ALMOST EVERYONE IN THE NASCENT PRO-NUCLEAR COMMUNITY IN THE 2000s had a conversion story. For some, it started with a YouTube video insisting that next-generation reactors would save the world. For others, it was sparked by a conversation with a respected scientist friend, or an op-ed by James Lovelock (one of the few who had apparently always favored nuclear). But it seems safe to say that nobody had a conversion story quite like that of Tom Blees.

Blees was an extremely tall autodidact who dropped out of college in 1971 because he thought he could learn more about his major, anthropology, by traveling the world than by sitting in a classroom. Thus began his adventures hitchhiking, riding freight trains, and "sky-hitching," a practice he may or may not have invented that involved showing up at airports where private planes took off and asking the passengers about to board them if he could tag along. (Ah, the '70s.) Eventually, he settled down—sort of: he married a French Canadian woman he'd met in Guatemala, had two children, and worked five weeks a year as a commercial fisherman in Alaska. He would earn enough in that season so that his family could live modestly for the rest of the year, in Florida or the West Indies or wherever they felt like.

In the late '90s, Blees and his wife had an idea for a nonprofit. They believed that expanding access to potable water would be the most effective way to save lives in Central America. They wanted to found an organization called the Aquarius Project to drill water wells in the region. While trying to raise money, Blees managed to land a meeting with someone from the Bill and Melinda Gates Foundation, which was then just getting off the ground. But she told him the foundation wouldn't go for it: "They're all techies," he recalls her saying. "Nobody wants to talk about diarrhea."[1]

The idea fizzled, but Blees decided to write a novel about a man who founds an organization called the Aquarius Project, which drills wells in poor countries throughout the world. The protagonist, Jack Lindsay, wins the Nobel Peace Prize for his efforts and, through a circuitous series of events, finds himself getting sworn in as president of the United States. Lindsay is an idealistic yet humble fellow with a sense of humor, principled and selfless, not to mention good-looking. As president, he boldly pursues his progressive agenda without regard for political consequences.

Perhaps Blees's motive was to get the word out, in an oblique way, about the importance of clean water; perhaps it was simply a means of wish fulfillment. Regardless, as he labored over the novel, Blees realized he had to write something about President Lindsay's energy policy. "So I was gonna write maybe a page," he told me. "But I wanted it to be interesting and kind of thought-provoking." One day he was listening to talk radio in the car, and someone mentioned a nuclear reactor that could burn nuclear waste as fuel. "And I thought, well, I wonder if this is a real thing. So I started looking into it. . . . And then I found out that something like this had been invented. And it was really hard to find out about it." After months of research, Blees started to piece together what had happened.

THE LOUD BANG WAS NOT REASSURING.

On April 3, 1986, a test was performed on an experimental nuclear reactor at a lab in Idaho. The control room was filled with about 30 people, some of them international observers. The objective was to see what would happen if the reactor core lost coolant—which is what had happened in Unit 2 at

Three Mile Island in March of 1979. But this was a very different type of reactor.

The lab was part of the network of more than a dozen National Laboratories that had grown out of the Manhattan Project after the Cold War to continue research on the atom as well as other scientific questions. One of the most prominent of these labs was Argonne, affiliated with the University of Chicago. The very first self-sustaining, controlled nuclear chain reaction had been carried out in 1942, in a squash court at the university, overseen by Enrico Fermi, the Italian physicist and Nobel laureate. Fermi became the first director of Argonne National Lab, a sprawling campus located in a suburb about 25 miles southwest of the city. In the late 1940s, as the United States was gearing up to develop a civilian nuclear reactor program, the Atomic Energy Commission sought out a good place to test reactors: somewhere much farther from an urban center than Argonne. They landed on an 890-mile expanse in the high desert in southeastern Idaho. The complex there became known as Argonne-West (it was later renamed Idaho National Lab).

From the beginning, it was known that there were a variety of possible designs for nuclear reactors. In what were called "thermal" reactors, the fission reaction would be moderated by a substance such as water. But Fermi had proposed another idea that sounded almost magical: If the neutrons were not moderated but permitted to move as fast as possible with high energy, the reactor would actually produce—or "breed"—more fuel than it consumed.[2] More neutrons would be released; when the excess neutrons collided with uranium-238 atoms (which could not serve as fuel), these atoms would be converted into plutonium-239 (which could). This concept was known as a "fast reactor" or a "breeder reactor."

At the testing station in the Idaho desert, the first reactor to ever produce electricity was, in fact, a breeder reactor. At 1:23 p.m. on December 20, 1951, Experimental Breeder Reactor No. 1 (EBR-1) powered four lightbulbs.[3] During this period, though, the famously abrasive and tireless Naval Admiral Hyman Rickover was overseeing the development of a thermal reactor, moderated by water, to power submarines for long distances. (It was known as a "light-water" reactor, to distinguish it from "heavy water," which contains more than the natural proportion of heavy hydrogen atoms; heavy water can also serve as a moderator.) Rickover succeeded in 1953, and the light-water reactor ended up becoming the model for all of the nation's civilian nuclear plants. Research on the fast reactor continued with another facility, Experimental Breeder Reactor-2, housed within a large silver dome. But not long after it launched, in 1964, this line of research was terminated.

In the early days, Argonne was an exhilarating place to work for scientists who were fortunate enough to land jobs there. They were given wide latitude and ample funding to pursue their research, and, at the time, nuclear physicists were widely respected, even revered. Charles E. Till, a Canadian nuclear physicist who was hired in 1963, wrote that working there was like being in a "minor aristocracy." As he was driving home from work one day, he thought, "Nobody tells me what I have to do. I am doing exactly what I know how to do. I am given all the equipment and all the help I need to do it. And I am very well paid. What a life this is."[4]

By the '80s, however, the shine had come off the field. Nuclear physicists were no longer treated like rock stars. And at the National Labs, there had been ups and downs related to funding, bureaucracy, and leadership. But research continued, and several breakthroughs had occurred that were relevant to fast reactors. Some scientists, including Till, thought the time was ripe for a fast reactor revival. Till directed the resulting project, with Yoon Il Chang, a Korean nuclear engineer, as his deputy. They called their concept the Integral Fast Reactor (IFR). As Till and Chang later wrote in a book they coauthored, *Plentiful Energy*, "Its purpose was the development of a massive new long-term energy source, capable of meeting the nation's electrical energy needs in any amount, and for as long as it is needed—forever, if necessary."[5]

Like any breeder reactor, the IFR would create more fuel than it consumed. But its innovative features went beyond that. In fact, Till and Chang deliberately tried to address all of the major concerns of anti-nuclear activists. Waste? Most of the waste would go directly back into the reactor and serve as additional fuel. There would eventually be a waste product, but a very small volume that would lose its radioactivity in a far shorter period than the spent fuel from conventional light-water reactors. Proliferation? They developed a new technique for reprocessing the waste, in which the plutonium would be mixed up with other elements, making it harder to divert into weapons production. Accidents? It would boast "passive safety features," meaning that because of the inherent properties of the materials, it would automatically shut down in conditions that could lead to a meltdown, without human intervention. Check, check, check.

That was the vision, anyway. A great deal of work would obviously need to be done to make it a reality. And before April 3, 1986, no one knew for sure how the reactor would perform under accident conditions. By that time, a prototype had been built in the old EBR-2, the reactor with the silver dome. The 30-odd observers, mostly European scientists who were also interested in pursuing fast reactors, gathered in the control room, while most of the Argonne staffers stood in the back. A lit-up red sign read "REACTOR ON," and a dark-haired man in glasses intoned, "T minus one minute" over a loudspeaker. Then the flow of coolant—liquid metal rather than water—to the reactor core was shut off. An indicator showed the temperature of the core rising past 1,100 degrees on a red digital sign.[6]

That's when the loud bang happened. "People got surprised," Chang told me. "That's not something they expected to hear, a big bang." As Till described it in a television documentary, "these heads went straight up like so, and turned right around to see whether the Argonne people were running."[7] What had happened was that as the steam pressure increased, the safety valve in the turbine opened to release it, making the noise. Nothing had gone wrong. And within seconds, the temperature indicator began to go down: as a result of the excess heat, the core expanded, the uranium atoms spread apart, and the temperature began to drop. The reactor shut itself down. The visitors relaxed. The experiment was a success. They then performed another test, for a different kind of accident, called "loss of heat sink," which was also successful. A press release was sent out, but nobody paid attention.

In nuclear lore, there is a famous coincidence: that the accident at Three

Mile Island occurred merely 12 days after *The China Syndrome* hit theaters. This is seen by nuclear advocates as unfortunate, because it primed people to be that much more fearful of the meltdown and suspicious of the industry. But there's another coincidence, less widely noted even within the pro-nuclear world. Less than a month after the successful test of the Integral Fast Reactor at Idaho National Lab, news broke of the accident at Chernobyl. At Chernobyl, when the temperature had begun to rise, it had surged uncontrollably, the reactor core had melted down, and a huge radioactive cloud was released, making its way across Europe. (Unlike all American nuclear reactors, the one at Chernobyl did not even have a containment dome.)

After Chernobyl, a journalist at the *Wall Street Journal* remembered seeing the press release about the test in Idaho, and it did get some modest media coverage then. Some people thought the IFR might represent a promising new avenue for nuclear power. But others thought nuclear power should be abandoned altogether.

NATIONAL LABORATORY PROGRAMS RELY PRIMARILY ON FUNDING FROM Congress—a notoriously tricky position. And among lawmakers, attitudes toward the IFR program were always mixed. After initial bipartisan support for nuclear power in the postwar period, opinions had largely polarized. The anti-nuclear movement, of course, had been associated with the New Left and the environmental movement, so many Democrats (not all) turned against nuclear. Republicans, however, remained generally supportive. It fit in with their worldview: pro-industry, prioritizing energy security above environmental protection. The IFR program was launched when President Reagan was in office, and under President George H. W. Bush, it continued to receive funding, although some of the votes in Congress were close.

By 1993, Till and Chang estimated that they would have the major kinks ironed out within a few years. But on February 17 of that year, President Bill Clinton, in his first State of the Union Address, announced, "We are eliminating programs that are no longer needed, such as nuclear power research and development." His proposed budget cut funding for the project.

(Funding to that point had been roughly $100 million a year, for a total of about $1 billion.)[8]

The debate in Congress was heated. Democratic representatives from Illinois supported the project, which provided hundreds of jobs for constituents who were working at Argonne. The two major points of contention involved fiscal responsibility and proliferation risk. In both cases, there were basic disputes about the facts. In terms of funding, the amount of money needed to shut down the program, according to proponents, was almost as much as, or even more than, the amount that would be needed to complete it. As for proliferation, reprocessing waste involves turning it into a form that is closer to weapons-usable. Although proponents pointed out that the new reprocessing technique used for the IFR was designed to reduce the risk—by keeping plutonium with several other elements—opponents still considered it unacceptably dangerous. John Kerry, who led the fight against the program in the Senate, said that his colleagues on the other side were pursuing an "irresponsible effort for fundamental pork-barrel purposes."[9] Till and Chang believed that the opponents were driven simply by anti-nuclear ideology. Although global warming was on Congress's radar by this point, it came up only rarely in the discussions.

The House followed Clinton's lead, cutting funding. Although the Senate ended up narrowly upholding it, in the subsequent conference to reconcile the two bills, the funding was eliminated. "The world is going to be a worse place 30 years from now without this technology," said an Argonne spokesman.[10]

Tom Blees had not been paying attention when this drama was playing out, but about seven years later, he began calling the PR guy at what was by this time called Idaho National Lab, who eventually admitted that he had been instructed not to publicize the project. (When I asked the current PR person at INL about this, she wrote in an email, "There is no way for us to confirm this claim.") Blees managed to reach Till and Chang and struck up a correspondence with them. He was easily convinced that the IFR could have been a virtual panacea and that it had been squashed for absurd political reasons. Blees wanted to write a book about what had happened, and Till and Chang wanted to make sure he grasped the technical details. They ended up tutoring him, mostly via email, for eight years.

In 2008, Blees, who was then living in Davis, California, self-published a nonfiction polemic, *Prescription for the Planet.* It was largely about the IFR, recounting the funding saga and making the case that it should be resurrected. The cover featured a transparent bottle of pills, some decorated with the symbol for the atom, others with lightning bolt icons. In the book, he coined the pejorative term "environists," to mean "Environmentalists in whom the 'mental' portion is substantially inoperative."[11]

A couple of months before his book came out, he decided to email the world's most prominent climate scientist: James Hansen. Blees wrote, in his recollection, "You're always saying what a crisis climate change is, but you never say what to do about it. So I wrote a book about what to do about it. Would you like to see an advance copy?" Hansen received hundreds of emails every day, but something about this one caught his attention. Blees had no credentials, not even a college degree, not even a publisher for his book, but apparently Hansen was indifferent to such niceties. He replied and said he was in Europe, but told Blees to send it to New York; he'd take a look when he returned.

"And I figured I'd never hear from him again," Blees told me. "About two weeks later, he calls me on the phone and he's really excited. Because he'd been trying for years to figure out how to run modern societies with renewables and had come to the conclusion that it was impossible. And he was really despairing. And when he read my book, he realized that this could be done, and he was really jazzed about it."

Hansen remembers it slightly differently. "My interest in nuclear power grew gradually," he wrote to me in an email. He told me that in the '70s, he had attended a talk by Amory Lovins, a well-known physicist and energy expert who was associated with Friends of the Earth. In a famous 1976 *Foreign Affairs* article, "Energy Strategy: The Road Not Taken?", Lovins laid out a sophisticated vision of the energy future. He distinguished between the "hard path" of the existing energy system and the "soft path" of increased energy efficiency and development of renewable energy sources. One major difference was that the "hard path" consisted of large, centralized power plants, whereas the "soft path" would comprise small-scale, distributed energy sources, such as rooftop solar. Another distinction contrasted the finite resources of oil, gas, coal, and uranium with the inexhaustible flow of energy from the sun and wind: the latter, he wrote, was "energy income," while the former was "depletable energy capital."[12] When Hansen heard Lovins speak, he thought his argument "made sense," he wrote me. But then he started to look more into energy issues in the late '90s and early 2000s. "I began making graphs for comparison with the scenarios that Lovins had painted for phasing down carbon emissions, and found that, although we were making progress with energy efficiency, the 'soft' renewables that he was counting on were stagnant as a small percent of total energy." He also began to talk with people who ran utilities and were responsible for keeping the lights on. They unanimously told him that using energy more efficiently and relying on renewables would not be enough, at least not anytime soon.

So he had already begun to ponder nuclear when he heard from Blees, but it is clear that Blees's book had a strong effect on him. At the time, Hansen sent out frequent messages to a global email list. On August 4, 2008, as part of a long "trip report" on his travels in Germany, the UK, and Japan, he wrote, "I read a draft of 'Prescription for the Planet' by Tom Blees, which I highly recommend." He reviewed all of the claimed advantages of the IFR, then concluded, "Some of the anti-nukes are friends, concerned about climate change, and clearly good people. Yet I suspect that their 'success' (in blocking nuclear R&D) is actually making things more dangerous for all of us and for the planet."

He found the argument about proliferation risk dubious. "Was it thought that nuclear technology would be eliminated from Earth, and thus the world would become a safer place?? Not very plausible—as Blees points out, several other countries are building or making plans to build fast reactors. By opting out of the technology, the U.S. loses the ability to influence IFR standards and controls, with no realistic hope of getting the rest of the world to eschew breeder reactors."[13]

Hansen also plugged *Prescription for the Planet* on Charlie Rose's television show and in his 2009 book *Storms of My Grandchildren*. As a result of this attention, Blees's life completely changed. He began hearing from government officials, scientists, and engineers from all over the world. He also started a Google Group where Hansen, Yoon Chang, and other scientists and engineers geeked out about the IFR and other nuclear technologies. It was one of the earliest gathering places for the budding pro-nuclear community. Although the IFR had been killed, the idea would go on to have a remarkable afterlife.

In *Storms of my Grandchildren*, Hansen described what happened when he started to speak publicly about nuclear, saying that the U.S. should, with urgency, resume work on the abandoned project and build a test fast-reactor power plant. He was inundated with messages cautioning against the pursuit of advanced nuclear. "Mostly it was friendly advice—after all, they agreed with my climate concerns—but they invariably directed me to one or more of a handful of nuclear experts," he wrote,[14] including Amory Lovins and representatives of the Natural Resources Defense Council and the Union of Concerned Scientists. As he saw it, they exercised a disproportionate influence, stifling discussions about a possible future for the peaceful atom.

"That's what began to make me a bit angry," he wrote. His anger derived from the same source that had driven anti-nuclear activists: fear for the future and for his progeny. He felt that a small group of dogmatic anti-nuclear advocates was taking off the table an important option for staving off catastrophe. "Do these people have the right to, in effect, make a decision that may determine the fate of my grandchildren?"[15]

Blees, Hansen, and others in their camp believed that climate change posed an existential threat. But they took for granted that energy-hungry modern society would continue on its path; unlike some Sierra Club members, they did not see dramatic reductions in energy consumption as realistic, nor as necessarily desirable. Given that assumption, they concluded that renewables, while they might hold great promise, were not up to the task of fully replacing fossil fuels. That meant nuclear had to play a critical role. Existing nuclear power had real problems, but with innovative technology—what Blees dubbed "newclear"[16]—those problems could be surmounted.

In December 2008, when President Obama was about to take office, Hansen wrote him and Michelle Obama an open letter calling for a three-pronged solution to deal with climate change: a carbon tax (unlike Shellenberger and Nordhaus, he would remain a staunch supporter of this policy); the phasing out of all coal-fired plants; and "Urgent R&D on 4th generation nuclear power with international cooperation." He warned: "The danger is that the minority of vehement anti-nuclear 'environmentalists' could cause development of advanced safe nuclear power to be slowed such that utilities are forced to continue coal-burning in order to keep the lights on. That is a prescription for disaster."[17]

BLEES DID END UP FINISHING HIS NOVEL, TITLED *EMERGENCY SESSION*, though it was never published. He sent me the 754-page manuscript. The section on his fictional president's energy policy, of course, was ultimately much longer than a page. In one scene, President Lindsay, having also discovered the history of the IFR, holds a press conference. "We can't let the fear associated with words like 'nuclear' and 'radioactivity' prevent us from making responsible decisions," he tells the reporters. "We have to reach beyond the realm of demagoguery and grasp the opportunity these technologies present to us. Global warming won't wait for us if we insist on sabotaging realistic solutions to these problems."

Slowly, impressed by President Lindsay's integrity and honesty, members of Congress start to come around. "And through it all, President Lindsay remained above the normal political fray as he pursued his own agenda, charting his course with a confidence that brooked no doubts that sufficient political support would materialize as needed."

Although the reality did not quite live up to this fantasy, an endorsement from the world's foremost climate scientist was not bad.

CHAPTER EIGHT

"WE ARE AS GODS AND *HAVE* TO GET GOOD AT IT"

On March 11, 2011, a 9.0 magnitude earthquake off the coast of Japan triggered an immense tsunami, which inundated a large area of northeastern Japan.[1] More than 15,000 people were immediately killed, and more than 6,000 were injured. The tsunami also damaged the electric power supply lines to the Fukushima Daiichi Nuclear Power Station, leading to a loss of the cooling function in three reactor units and the pools where spent fuel was stored. Reactor cores in the three units overheated, and radioactive isotopes were released into the air and the sea. To escape the radiation, thousands of people were evacuated from the area.

This disaster, the first major nuclear accident since Chernobyl in 1986, sparked a new wave of anti-nuclear sentiment around the world. Japan temporarily shuttered all of its plants. At an event days after the tsunami, German chancellor Angela Merkel, who had formerly supported nuclear power, surprised anti-nuclear activists, who were in the audience booing her. She said that the "alarming events" in Fukushima had "changed a few things," and announced her intent to phase out the country's nuclear plants by 2020.

"If we can reach this goal sooner, all the better."[2] Other government leaders were grappling with how to respond.

The incident also posed a conundrum—albeit with much lower stakes—for documentary filmmaker Robert Stone. Stone's debut, released in 1988, had been a devastating Oscar-nominated documentary called *Radio Bikini* about the U.S. military's nuclear testing program in the Marshall Islands. It intersperses footage of the run-up to the test on Bikini Atoll and the explosion itself with two heart-wrenching interviews: one with the leader of the people who lived there and were forced from their homes; the other with a haunted American veteran describing witnessing the test and receiving assurances from his superiors that he was not at risk. At the end of the film, the camera pulls back to reveal that he is missing his legs, which had to be amputated due to radiation exposure. If an anti-nuclear march had passed by my house while I was watching that movie, I would have run outside to join it.

In 2009, Stone had released a documentary called *Earth Days*, which featured prominent environmentalists reflecting on their efforts of the '60s and '70s. The talking heads included Hunter Lovins, a renewable energy advocate (and former wife of Amory); Denis Hayes, who organized the first Earth Day; and Stewart Brand, founder of the hippie bible *Whole Earth Catalog*. Brand, in particular, was a kind of guiding light to Stone and to many others. Steve Jobs cited him as an inspiration, calling the *Whole Earth Catalog* a precursor to Google. Ted Nordhaus and Michael Shellenberger quoted him in their 2007 book *Break Through*: "We are as gods and might as well get good at it."[3]

While Stone was making *Earth Days*, Brand was working on a book that would make the case for nuclear power, a turnaround for him. Over the years, Stone had also become more open to the idea. The subject didn't come up in that film, but when it was screened at the Sundance Film Festival, the first question from the audience was, "What is your opinion on nuclear power?" (This seems to be a theme.) Stone answered, as he later recalled, "I have to tell you my views on that subject have really changed. But I have two people here who I think can really address that." First Hayes spoke, insisting that nuclear was not the answer. Then Brand came out and gave his pro-nuclear spiel. "The room went fucking crazy," Stone told me—as if the audience had suddenly realized, "'Oh, we can talk about this? Oh, thank God.'"

I met Stone for dinner at a French restaurant in midtown Manhattan in the summer of 2022, and amid the din, he told me he didn't see support for civilian nuclear energy as at all inconsistent with his first film; conflating weapons with energy production was part of the problem, he said. So he decided to make his next documentary about the peaceful atom. He'd come to believe that "the environmental movement has been adamantly opposed to the most viable solution we have to the greatest environmental catastrophe we've ever faced," and he thought that apparent contradiction "could be a good premise for a movie," he told me. He had raised money for the project and had started shooting when he heard the news from Fukushima.

THE BOOK STEWART BRAND PUBLISHED IN 2009, *WHOLE EARTH DISCIPLINE: An Ecopragmatist Manifesto*, was obviously a callback to his earlier work, but in many ways, it represented a rupture with it.

The *Whole Earth Catalog* was an unclassifiable publication whose target audience was the growing numbers of young Americans who were making a go of living in rural communes. Starting in 1968, Brand and his wife, Lois Jennings, assembled the issues in a disheveled office in Menlo Park. The catalog's tagline was "Access to Tools," and it consisted mostly of product reviews. You could not order items from the catalog, but you could learn about them and where to find them. The featured items included weaving kits, potters' wheels, bamboo flutes, calculators, camping gear, geodesic domes, windmills, solar panels, and plenty of books.[4] So, on the one hand, the catalog was essentially facilitating shopping. On the other hand, the goal of the purchases was, somewhat paradoxically, to liberate people to live outside of consumer society. The particular items highlighted were those that could enable communards to live in a more DIY fashion. It was fostering a new network of buyers and sellers beyond the mainstream. Brand was never against commerce, or, unlike some of his contemporaries, even capitalism.

But he was an environmentalist, and those values informed the catalog. Solar panels and windmills were not only green but seemed to go perfectly with the publication's ethos, freeing people to live off the grid and generate their own energy. In contrast to some environmentalists, Brand was not anti-tech. His bent was more curiosity and openness to new technology. As he would later put it, he had "an engineer's bias, which sees everything in terms of solving design problems."[5]

As for nuclear energy, he had always subscribed to the near-consensus in his cohort that the problems with nuclear reactors, especially the creation of highly radioactive waste, were deal-breakers. The gargantuan centralized plants were also the opposite of the kind of small-scale distributed energy and individually controlled tools celebrated in the *Whole Earth Catalog*. Like James Hansen, he was influenced by Amory Lovins, who became a friend.

But by the early 2000s, Brand had started to question his previous convictions. By this time, he was living with his second wife, Ryan Phelan, on a houseboat in Sausalito, in the Bay Area. He was involved in two organizations, the Global Business Network, a consulting firm; and the Long Now Foundation, which was dedicated to long-term thinking about humanity's challenges. Through his work, he began to learn more about the risks of climate change and—prompted by a visit to Yucca Mountain—began to think more about nuclear waste.

In 1982, the Nuclear Waste Policy Act assigned the federal government the task of constructing and operating waste disposal facilities. According to the law, the Department of Energy was required to establish two deep geological repositories for spent nuclear fuel. The idea was that one would be in the eastern part of the country, the other in the West, to distribute the burden fairly. In 1987, however, amendments to this law identified only a single location: Yucca Mountain. It was deemed conducive to digging tunnels where canisters of radioactive waste could be placed about 350 meters below the ground surface. It was in a remote area, and risks were considered low. By the time Brand visited, billions of dollars had been sunk into the project.

Brand learned that more than 90 percent of the energy in spent fuel was untapped, with the potential to be recycled.[6] He also learned about the way it was currently managed. What Mothers for Peace had been told was correct: it was essentially the same stuff as fallout—that is, radioactive fission products. But unlike fallout, it was isolated, contained, and monitored. This was also in contrast to the fossil fuel pollution that was spewed willy-nilly into the air. Brand began to wonder if he had been thinking about nuclear power all wrong. "Waste disposal no longer looked like a cosmic-level problem, and carbon-free energy from nuclear looked like a major solution in light of growing worries about climate change," he wrote in *Whole Earth Discipline*. "My opinion on nuclear had flipped from anti to pro. The question I ask myself now is, What took me so long?"[7]

He cited a number of others who had gone on similar journeys, such as Hansen and Gwyneth Cravens, the former *New Yorker* editor who had published *Power to Save the World: The Truth about Nuclear Energy* in 2007. And he discussed a contingent of environmentalists he dubbed "reluctant tolerators," including Bill McKibben, who had written, "It's not enough for greens to say that nuclear power is risky and comes with consequences; everything comes with consequences."[8]

In the book, Brand laid out a comprehensive case for nuclear power. Like Blees and Hansen, he was enthusiastic about the prospects of next-generation nuclear. But while they (at the time) expressed reservations about existing reactors, Brand was making his case more generally in favor of nuclear, including the current facilities. He argued that fear of nuclear was disproportionate to the actual risk, mentioning the disaster in Bhopal in 1984, when a gas leak from a pesticide plant immediately killed at least 3,800 residents.[9] Although this was far more than the initial death toll at Chernobyl two years later—a total of 30 within three months of the accident[10]—the name "Bhopal" had hardly lodged in the public imagination in the same way as Chernobyl had.

And he wasn't just arguing that renewables were not capable of providing all of our energy—he was also pointing out other problems with them. The advantages of nuclear have to do with a key feature of uranium: its energy density. That means that nuclear plants need less raw material and less land than renewables to produce the same amount of energy. According to one of

Brand's sources, a 1,000-megawatt nuclear plant occupied a third of a square mile, compared with more than 200 square miles for a wind farm and more than 50 square miles for a solar array.[11] Because of the land requirements for renewables, one expert Brand quoted, Jesse Ausubel, went so far as to write, "Nuclear energy is green. Renewables are not green."[12]

Climate change, Brand suggested, wouldn't allow for the luxury of ideology. "Forty years ago, I started the *Whole Earth Catalog* with the words, 'We *are* as gods, and might as well get good at it.' Those were innocent times. New situation, new motto. 'We are as gods and *have* to get good at it.'"[13]

The book included other "environmental 'heresies,'" as Brand would put it in a TED Talk related to the book in June 2009. For example, he made the case for the benefits of large, dense cities. Today, the green virtues of cities are widely appreciated, but historically, environmentalists had romanticized small villages and the countryside—as the back-to-the-land movement reminds us—and Brand was an early proponent of the pro-urban view. (The YIMBY movement that would emerge a few years later aligned with this perspective.) He also argued for genetically engineered crops, excoriating groups like Greenpeace for opposing this technology, which he said could not only save lives in poor countries, but could grow the same amount of food on less land than conventional crops.

One common thread between the different phenomena he was endorsing was density: when people conglomerated in dense urban areas, that led to economies of scale and left rural areas to regenerate. The same was true when crops were genetically modified to grow more food on less land. Ditto nuclear power. Another theme was that the "natural" wasn't necessarily good for nature. Environmentalists had always tended to prize the countryside, organic farming, and energy sources that gently harnessed the sun and wind, all of which seemed to honor nature rather than try to alter it. Brand was arguing that sometimes it was the technologies that seemed to subvert nature the most—splitting atoms, splicing genes—that would end up saving the planet.

Whereas at least some strands of old-school environmentalism had a distinctly antihuman perspective, Brand's ecopragmatism sought to give equal weight to human flourishing. He noted that climate change had forced other environmentalists to think more about protecting humanity, too. "Greens are no longer strictly the defenders of natural systems against the incursions of civilization; now they're the defenders of civilization as well."[14]

In advance of publication, the book was teased as a major new work. But when it came out, it didn't exactly make a big splash. It was hardly even reviewed in the United States. Brand blamed his old friend Amory Lovins, who wrote an early, withering review in *Grist*, refuting all of his arguments about nuclear energy.[15] In Brand's view, this may have scared off other potential reviewers.[16] Among the small number of readers, though, were Nordhaus and Shellenberger. They found it revelatory. "That book is the thing that allows us all to see this different environmental politics," Nordhaus told me.

It got more attention in the UK, but the attention was far from uniformly positive. Brand appeared on a British television program called "What the Green Movement Got Wrong," along with Mark Lynas, a well-known British environmental activist who had also become pro-nuclear. Lynas got intense blowback for this from his former allies in green circles. As one blogger put it, "That guy went from hero to zero in six seconds[.]"[17] After a screening of the show, Brand and Lynas participated in a live studio debate. On the other side were environmentalists including George Monbiot, one of Lynas's close friends. It turned into what Brand's biographer calls "the worst night of their lives."[18] Lynas agrees. "I remember just feeling hounded, like literally I wanted to leave with a bag over my head," he told me. "This intense feeling of, I don't know—it wasn't paranoia, because those people actually did hate me."

WHEN STONE HATCHED THE IDEA OF MAKING HIS PRO-NUCLEAR DOCUMENTARY, he wanted to capitalize on the "power of the convert," he told me, featuring five advocates who had changed their minds. One of them, of course, would be Brand. He also recruited Mark Lynas; Gwyneth Cravens, the former *New Yorker* editor; Richard Rhodes, the Pulitzer Prize–winning author; and, finally, Michael Shellenberger.

When the accident at Fukushima happened, it forced Stone—as well as his subjects and anyone else who had begun to warm up to nuclear power—to, at the very least, do some hard thinking. But ultimately, it didn't slow the growth of the embryonic pro-nuclear movement; if anything, counterintuitively, it may have accelerated it.

Shortly after the incident, on March 21, George Monbiot, who had sparred with Brand and Lynas on live TV, wrote a column in the *Guardian*. "You will not be surprised to hear that the events in Japan have changed my view of nuclear power," he began. "You will be surprised to hear how they have changed it." He explained that while the accident had exposed familiar problems with the nuclear industry, no one had apparently yet received a fatal dose of radiation. "Atomic energy has just been subjected to one of the harshest of possible tests, and the impact on people and the planet has been small," he wrote. "The crisis at Fukushima has converted me to the cause of nuclear power."[19] On September 7, 2012, Nordhaus and Shellenberger published, along with their colleague Jessica Lovering, a piece in *Foreign Policy*. "Arguably," they observed, "the biggest impact of Fukushima on the nuclear debate, ironically, has been to force a growing number of pro-nuclear environmentalists out of the closet, including us."[20] (It's worth questioning whether, particularly in the case of Monbiot's piece, published 10 days after the accident, enough time had elapsed to assess its seriousness, since one of the biggest concerns about nuclear meltdowns is their long-term health effects. In the years to come, debate would continue about the nature and degree of severity of the accident, a topic to which we will return.)

Stone, too, concluded that the meltdowns, while unfortunate, were nothing close to an existential threat on the order of climate change. The accident did change how he approached his film: he and Lynas visited the exclusion zone around the plant and shot footage there. ("We had to sneak in," he told me. "We pretended we were international radiation observers. They eventually threw us out.") But he never seriously considered abandoning the project.

The movie, *Pandora's Promise*, was released in 2013. It sent a message to a certain type of person—the type of person who read the *New Yorker* and sought out Pulitzer Prize–winning books—that it was okay to be pro-nuclear. It was, so to speak, safe. These smart people who shared your values and had done their homework had changed their minds, and you could, too. Shellenberger was perhaps more controversial than the others, but a) most people probably didn't know who he was (or that he was considered controversial); and b) he was still, at that time, operating within the bounds of the mainstream. The last words of the movie go to him: "You actually do feel like this is the beginning of a movement."

Pandora's Promise had modest success at the box office, but it ended up being pivotal for the pro-nuclear movement in other ways. One had to do with funding, as the film helped the nuclear advocates attract friends with money and influence.

Among them was a venture capitalist named Jim Swartz, a major funder of documentaries, who signed on early as a producer for *Pandora's Promise.* One day, in search of more funding, he called his friend Ray Rothrock, a Texas native who had worked as a nuclear engineer before moving to the tech industry and then also becoming a venture capitalist. Rothrock signed on as a producer, too. Swartz then hosted a gathering at a conference room in his office in Palo Alto, with entrepreneurs, VCs, and philanthropists in attendance. Nordhaus and Shellenberger were also there, and Stone screened some early footage. "And literally, that night, one night, it sounds like church, but we passed a hat and we raised, like, a million and a half dollars," Rothrock told me. Contributors included Google researcher Ross Koningstein and tech entrepreneur Steve Kirsch.

Every June, the Breakthrough Institute held a conference at a swanky resort in Sausalito, just across the Golden Gate Bridge from San Francisco. Those donors attended the next one, as did philanthropists Rachel and Roland Pritzker, heirs to a hotel fortune, who had also chipped in to support the film. Each of them—Swartz, Rothrock, Koningstein, Kirsch, and the Pritzkers—contributed $50,000 to support research and advocacy for nuclear energy at the Breakthrough Institute. One of the first fruits of the funding was a 2013 report called "How to Make Nuclear Cheap." "That initial funding for BTI's nuclear work was, to my knowledge, the first significant philanthropic funding for pro-nuclear advocacy ever," Nordhaus wrote to me in an email. It "was really what built the entire field," he wrote. "There was really no NGO advocacy and no funding before Pandora's Promise, which provided a high profile event around which to rally funding and organizing."

The making of the movie also created the conditions for a small posse of nuclear advocates to coalesce. The five featured advocates spent time together promoting it, hanging out at Sundance and other festivals. There were a number of screenings around the country with discussions afterward. Rothrock, who attended some of them, told me, "Invariably, the people in the room would go, 'This is a great movie. Now what? What's the call to action?'"

To try to answer this question, Rothrock held a two-day meeting at his home in Portola Valley, a small, exclusive town in the Bay Area. More than a dozen people sat around his living room, whose floor-to-ceiling windows looked out on the Santa Cruz Mountains. Attendees included Nordhaus, Shellenberger, and Rachel Pritzker; Pritzker's philanthropic advisor, Mike Berkowitz; and staff members of Clean Air Task Force (CATF), a Boston-based organization, and Third Way, a center-left D.C.-based think tank. They continued the conversations over dinner at a nearby restaurant, the Parkside Grille. Berkowitz told me the sentiment was, "We have to do something here. We've identified a key solution, part of the toolbox for addressing climate, and, like, nobody's doing anything on any of this." They debated questions such as how best to allocate resources—to lobby inside the Beltway, or try to change public opinion first? To save threatened plants or promote new advanced nuclear projects?

In the coming years, Third Way, CATF, and the Breakthrough Institute would assume complementary roles. Third Way did most of the behind-the-scenes work, cultivating relationships with lawmakers and their staffs and briefing them on advanced nuclear. Many of those conversations were based on research from CATF (which was more academic) and the Breakthrough Institute (more in the public-intellectual vein). CATF would release wonky white papers on, for example, how to create a more effective licensing pathway for the NRC, Rachel Pritzker told me. (She helped fund all three organizations.) Then "Third Way could sort of take it around to the right offices. And Breakthrough would do the big, high-level case making and research." It was, says Berkowitz, "this confluence of civil society actors, the first time you ever have a nonindustry pro-nuclear kind of movement."

How to characterize this incipient movement? For the most part, the people involved were politically center-left or not far from there and were focused on climate. They were inclined to respect what they saw as an underappreciated consensus among experts about the need for nuclear and were prepared to buck the conventional wisdom among environmentalists. Indeed, being pro-nuclear seemed subversive, and for some, that might have been part of the appeal. Josh Freed, leader of Third Way's Climate and Energy Program, who called in to the meeting at Rothrock's house, told me that he had been part of the D.C. punk scene, attending Fugazi shows in small clubs and playing bass in his own band. He said that being in the early pro-nuclear community reminded him of being a part of that scene.

That is not to say that these advocates were insincere, or to deny that in some cases it required courage to publicly support nuclear; it is only to acknowledge that the frisson of feeling like a maverick has always been one element of the pro-nuclear movement. Yet, while their numbers were few, this was not your typical "small group of thoughtful, committed citizens," in Margaret Mead's (possibly apocryphal) famous quote, who go on to change the world. It was very wealthy donors funding elite think tanks. (In terms of industry funding, CATF and the Breakthrough Institute say they do not accept money from energy interests, while Third Way does not discuss its donors.) They had connections, they had platforms, and they were making plans.

But even as this movement was cohering for the first time, fissures were starting to appear. Specifically, a schism was brewing that would split apart the Breakthrough Institute. And just as at the Sierra Club almost half a century earlier, the catalyst would be Diablo Canyon.

Nordhaus and Shellenberger founded the institute in 2007. Their offices were in the heart of downtown Oakland, on the eighth floor of a historic,

15-story Beaux Arts building on the corner of 14th Street and Broadway, right across from City Hall. At the think tank, they had continued to straddle the line between constructive criticism of and engagement with the constituencies they'd originally identified with—environmentalists, Democrats, progressives—and essentially trolling them. In 2011, David Roberts of *Grist* described the Breakthrough founders this way, dating back to "Death of Environmentalism": "If you cared to pick them out, there were good ideas about the need for technology innovation in clean energy. Those ideas were surrounded by a grand and elaborate exercise in myth-making that cast green groups as the fount of all error, a fusty and hidebound Establishment forever attempting to diminish and suppress the forces of the New Paradigm, led by, of course, S&N themselves."[21]

To their credit, though, they engaged in dialogue with those they disagreed with (even Roberts would later be a guest on a Breakthrough podcast). And they attracted a series of bright and idealistic youngsters who found their overall philosophy compelling and, depending on the person, either didn't mind their combativeness or were willing to overlook it. These included Tyler Norris, who would go on to work in Obama's Department of Energy; Jessica Lovering, who would later found a progressive pro-nuclear policy organization; Jesse Jenkins, who would become a Princeton professor and prominent energy expert; and Alex Trembath, who would stay at Breakthrough and eventually become deputy director. After work, the young staffers would go out together to local bars—Make Westing, known for its spicy habanero popcorn, or the Golden Bull, a music venue right next to the office building. Working for a heterodox environmental think tank in the thick of the Bay Area—home of the Sierra Club and the counterculture and still a hotbed of left-wing and environmental activism—could be stressful, and they needed to unwind.[22]

Since the mid-1990s, Shellenberger and Nordhaus had been virtually inseparable. Each was best man at the other's wedding. They coauthored dozens of articles and essays, as well as a book. For a time, they shared a single author bio. Although they were very different people, sometimes those who knew them had trouble figuring out how to distinguish between them, or who did exactly what in their partnership. Even as the young staffers were getting their own Twitter (later known as X) accounts, Nordhaus and Shellenberger tweeted only from the Breakthrough Institute account, in order to speak with one institutional voice. In retrospect, the nature of the partnership between the two cofounders was a bit surprising. It was unusual for any two writers to merge their voices in that way, and these two—especially Shellenberger—were not known for their small egos.

Alex Trembath dates the tension back to *Pandora's Promise*, when Shellenberger was featured on the big screen—possibly another way that film played a pivotal role. (Nordhaus concurs: "Michael always wanted to be famous. He always wanted to be on TV.") Not long afterward, Shellenberger was invited to debate Ralph Nader about nuclear power on *Crossfire*, the CNN show. Around the same time, he and Nordhaus also finally started their own, separate Twitter accounts. For the first time in his professional life, Shellenberger was getting a taste of what it was like to be a solo celebrity, not to share the spotlight with his best friend.[23]

Then, in 2015, Shellenberger started hearing rumors that Diablo Canyon might be shut down sooner than expected. Diablo had a special place in his heart. He remembered that when he was about 12, his sister had learned about the facility. She'd said, "Can you believe they're going to build a nuclear plant on a fault line in California?" He had an image of a mushroom cloud rising from the site and killing everyone in the state. Later, he visited the site and was impressed by the workforce and the splendor of the area surrounding it. "I was just completely smitten with this amazing plant," he told me. "It's just so beautiful. The intertidal areas were pristine precisely because they don't allow schoolchildren to stomp around on them." He thought, *This is everything we say we want.*

Even at that meeting at Rothrock's house, different priorities had emerged. In theory, they could have been complementary, as different people focused on different areas, and to some extent they were. But Shellenberger was diverging from his colleagues, becoming more critical of both renewables and advanced nuclear. He decided that the way forward was to focus on saving plants that were threatened with closure. To this end, he was seized with the conviction that the Breakthrough Institute needed to transform from a think tank into a grassroots advocacy group. The young staffers were energy and policy analysts, not activists, so they would have no place in this revamped organization he envisioned.

By this time, Nordhaus was used to dealing with Shellenberger's whims. Sometimes he would watch in amusement and other times he would say, "No, Michael, we're not going to do that." At the Breakthrough Institute's annual conference one year, Nordhaus roasted his best friend, albeit affectionately, noting that he was "really easy to make fun of": "There was the month back in the summer of 2007 that Michael spent stalking David Axelrod, so sure was he that the two of us had the answer to Obama's struggles early in his presidential campaign. Michael even bought a special cell phone, in anticipation that Obama himself would need a private line to call him for advice."[24]

At first the antics had been fun, and sometimes his ideas were inspired. But Nordhaus found it all increasingly exhausting. Other festering tensions between the two leaders came to a head, relating to Shellenberger's management style and his public persona, newly unbound by Twitter. In his tweets, Shellenberger had become too abrasive for Nordhaus's comfort, which was saying something.

It became clear that they couldn't continue working together. One day Nordhaus confided in his father, Bob, that he thought he was going to have to leave the Breakthrough Institute. His father—"God bless him," Nordhaus said—asked him, "Wait, why should *you* be the one to leave?" Shellenberger was the one who wanted to transform it into a different type of organization. In the end, Rachel Pritzker came in as a kind of peacemaker and agreed to put up some funding so that Shellenberger could start his own group. Shellenberger left the Breakthrough Institute; Nordhaus would lead it alone.[25] "There was a power struggle, and I lost," Shellenberger told me. His long partnership with Nordhaus came to an end, as did their friendship. Their break became known in their circle as "the divorce."

Shellenberger formed a small new nonprofit, called Environmental Progress, that would focus on advocacy to save nuclear plants, starting with

Diablo Canyon. He organized an open letter, signed by James Hansen, Stewart Brand, and nearly 60 others, objecting to the planned closure of the plant. "Diablo Canyon provided 22 percent of all the clean energy electricity generated in California in 2014," the letter read. "If closed, it will likely be replaced by natural gas and California's carbon emissions will increase the equivalent of adding nearly two million cars to the road."

Shellenberger also launched a website called "Save Diablo Canyon." Not long thereafter, he got an email from a woman named Heather Hoff.

CHAPTER NINE

THE DEAL

On January 31, 2012, at about 4:30 p.m., a sensor detected a leak at San Onofre Nuclear Generating Station, a nuclear plant located on the California coast midway between Los Angeles and San Diego. The facility, known as SONGS, had operated since 1968. It employed roughly 2,000 people and provided enough electricity to power 1.4 million homes. The giant domes of its two reactors were visible to drivers on Interstate 5, a heavily trafficked freeway, and to surfers who schlepped from all over the world to catch waves at famed beaches, such as Trestles and Cottons Point, nearby.

The utility that ran the plant, Southern California Edison, had recently replaced the reactors' steam generators, major parts consisting of thousands of metal tubes. When the leak was discovered in one of the generators—a small amount of radioactive water escaping from a tube—workers shut the reactor down within an hour. Two days later, Edison announced that the tubes in both generators were, mysteriously, showing extensive premature wear. The plant would remain offline indefinitely while Edison tried to determine the exact extent of the problem and what to do next.

When S. David Freeman, a retired energy expert, heard this news, he saw it as a sign. Just months earlier, Freeman had begun strategizing about how to bring the nuclear era in California to an end. He interpreted the leak—which, fortunately, was too miniscule to pose any danger to the public—as a message from the plant: "Shut me down for good before I kill you."[1]

Freeman was a legendary figure in the American electricity sector. In his numerous jobs leading public utilities—from the Tennessee Valley Authority to the Los Angeles Department of Water and Power—he was known for promoting energy efficiency and renewables and for killing nuclear plants. He was also a bon vivant, a thrice-married man who enjoyed a good scotch. He called himself "the green cowboy": the son of an umbrella repairman, he had grown up as one of the few Jews in Chattanooga, Tennessee. He spoke with a southern drawl and wore a cowboy hat, which was as much medical choice as fashion statement; his dermatologist had told him he needed to cover his balding head to protect it from the sun.[2]

He traced his approach to energy policy largely back to a single meeting he'd held when working as an energy advisor to President Lyndon B. Johnson. One day at the White House, he met with two women from New Hampshire who were opposed to plans for a nuclear plant near their homes. They made the case that measures such as home insulation and more efficient lighting and refrigerators would conserve more electricity than the plant would generate, while also saving money. Freeman, despite his expertise in energy, had never given much thought to efficiency. "I listened to them and examined their numbers, and as I did, I felt as if a light bulb, a very efficient one, went off in my head," he later wrote.[3]

From then on, energy efficiency became his North Star. This at times put him at odds with prevailing sentiments in the White House: he found himself up against a Cold War–era mentality that exuberant profligacy was the American way, in contrast to the drab austerity of the Soviet Union. At the same time, a focus on conserving resources was not entirely fringe; millions celebrated the first Earth Day in April 1970, with support from Senator Gaylord Nelson and other lawmakers. A few years later, Freeman was instrumental in formulating the landmark corporate average fuel economy (CAFE) standards for automobile efficiency, which became law in 1975. (He was close to Ted Nordhaus's father, Bob, who also worked on this legislation.

Ted remembers meeting Freeman multiple times as a child at the Aspen Institute's energy convenings.)

Freeman's "green turn" included an attraction to renewable energy. As he saw it, energy efficiency and renewables went hand in hand. He thought the government was perpetually embroiled in obscenely expensive technological boondoggles, spending billions to research fission breeder reactors as well as nuclear fusion, that never became reality. Instead, if the country cut down on its need for energy—which he thought it could do without seriously impairing quality of life—it might be able to rely largely on the relatively simple technologies of wind and solar.

Environmentalists generally shared his views, but, in his subsequent roles leading utilities, he was in a better position than most to do something about it. As chairman of the TVA, he halted construction of 8 of the 10 nuclear plants that had been underway when he'd joined.[4] He believed the projections used to justify the plants were wildly overblown, and then his energy efficiency program had further cut the need for electricity. When he later led the Sacramento Municipal Utility District, he and his partners launched a program to plant a million trees in 10 years, to provide shade and reduce the appetite for air-conditioning, and he oversaw the closure of the Rancho Seco nuclear plant near Sacramento. (This plant had a history of problems, and the decision to close it was made by a referendum on the local ballot.) For most of his career, Freeman's anti-nuclear views had been primarily a matter of cost, but a visit to Chernobyl after the 1986 accident had left him shaken and convinced him of its dangers as well.

By 2011, Freeman had retired and moved to Washington, D.C., to be near his grown children and his grandchildren. There, he became close friends with Damon Moglen, who directed the climate and energy work at Friends of the Earth. Together, they decided to try to phase out nuclear power in California. They both already opposed nuclear, and the accident at Fukushima in March of that year only galvanized them further. The disaster also directed intensive media scrutiny and public concern to the risks at California's two remaining plants, SONGS and Diablo Canyon, which, like the one in Fukushima, were located on the coast in seismically active areas.

"U.S. Nuclear Plants Have Same Risks, and Backups, as Japan Counterparts," read a March 13 headline in the *New York Times*.[5]

When the temporary shutdown of SONGS was announced, Freeman and Moglen were ready to act. Friends of the Earth hired an independent nuclear engineer named Arnie Gunderson to write a report about the problems. They pursued legal action with the NRC and the California Public Utilities Commission (CPUC) to delay or prevent the plant's reopening. Local activists who lived near the plant also stepped up, holding rallies where they chanted, "Shut it down! Shut it down!" Eventually, Edison leaders conceded that fixing the defective steam generators would not be viable. (They blamed Mitsubishi, which had manufactured them and failed to deliver a convincing plan to repair or replace them.) In June of 2013, the utility, citing the expense of the ongoing outage, announced that both units would be permanently shuttered.

Questions would swirl for years about the exact reasons behind the decision. The cost was clearly a factor, but it was a matter of debate whether Edison would have closed the plant in the absence of external pressure. Pro-nuclear activists would blame the anti-nuclear activists, and the anti-nuclear activists, including Freeman, would take credit. Freeman found the victory sweeter than his many professional accomplishments, because of the "David-versus-Goliath" aspect.[6] It was time to move on to Diablo Canyon.

It was no secret that Friends of the Earth wanted Diablo Canyon to close, too. The group had been born out of the conflict over the plant, when David Brower left the Sierra Club to found a new organization, in part due to all the heated disputes over whether to endorse the plant's construction. As Damon Moglen, Freeman's friend in D.C., told me, "Diablo Canyon is kind of hardwired into the psyche of Friends of the Earth."

As PG&E undertook the process of applying to renew the plant's NRC licenses, Friends of the Earth had a couple of arrows in its quiver, raising objections based on seismic safety and costs. David Brower had reportedly quipped, "Nuclear plants are incredibly complex technological devices for locating earthquake faults."[7] In the case of Diablo Canyon, that had seemed to be true—the Hosgri Fault had been discovered when the plant was under construction. But even Brower might not have anticipated that faults would continue to be discovered long after the plant began operating. In 2008, a geologist with the U.S. Geological Survey had discovered one 985 feet from Diablo Canyon's intake structure; it was eventually named the Shoreline Fault. Other faults—the San Simeon, San Luis Bay, and Los Osos Faults—had also been identified.[8]

In October of 2014, Friends of the Earth submitted comments to the NRC, requesting a hearing and seeking to intervene in the license renewal proceeding. "PG&E has not demonstrated that the plant can be safely shut down following an earthquake on one or more of these faults," they argued. Around the same time, the organization commissioned a prominent renewable energy advocate, V. John White, to conduct an economic analysis of PG&E's options. Looking at costs needed to continue the operation of the aging plant—including potentially expensive upgrades—as well as four different scenarios for renewables expansion and efficiency improvements, the report concluded that extending the life of the plant would cost at least $17 billion, while the renewables alternatives would range from about $12 billion to about $15 billion.[9]

For Friends of the Earth, as for most environmentalists, the vision for California's future was both nuclear-free and fossil-free. The dream of an all-renewables future had an obvious appeal. Solar panels would blanket rooftops, and wind turbines would rise from fields, generating electricity in a way that was more in tune with nature. This vision was also more inherently democratic: small-scale and dispersed, solar and wind installations could work at the community and even the household level. Not only that, the technologies were straightforward and easy to understand; they did not entail the almost unfathomable complexity, or require the advanced expertise, that operating a nuclear power plant did, with its backups and redundancies, its seismic analyses and armed security guards. As Amory Lovins had written, the "soft path" would "offer a good prospect of stability under a wide range of conditions, foreseen or not. The hard path, however, is brittle; it must fail, with widespread and serious disruption, if any of its exacting technical and social conditions is not satisfied continuously and indefinitely."[10]

The soft path was the one that Friends of the Earth championed. And even in the preceding few years—since Hansen and Blees had concluded that renewables were insufficient—stunning progress had been occurring in renewables development. Policies like renewable portfolio standards and tax rebates encouraged installations, which led to falling prices, which in turn encouraged more installations, in a virtuous circle. In 2007, for example, there was virtually no solar power produced in the U.S.; in 2015, solar accounted for 29.5 percent of the country's new generating capacity, exceeding new natural gas for the first time.[11] To be sure, wind and solar still only accounted for a small fraction of total net generation: 4.6 percent from wind and just 1.1 percent from solar (as well as 0.4 percent from geothermal), compared with 19.4 percent from nuclear. But the share of renewables was growing rapidly and was projected to continue rising.[12]

Freeman, while determined to close Diablo Canyon, wanted to be strategic. Although he had counted the closure of SONGS as a victory, he soon realized that it had come with some unfortunate consequences. Most of the plant's roughly 2,000 employees had been immediately laid off. The 2,200 megawatts of electricity produced by the plant had been withdrawn from the grid without any backup plan and was largely replaced by natural gas.[13] By no means did Freeman regret his pressure campaign to close the plant—he thought it was patently unsafe to continue running it—but he recognized that the abruptness of the shutdown was bad for workers and for the climate.

In Diablo Canyon, Freeman saw an opportunity to do it right. He sought to plan for a gradual, "orderly" transition, one that would both offer a softer landing for the workforce and allow time to build up renewable resources to replace the lost electricity. In late 2015, he called his longtime friend Ralph Cavanagh, a senior leader at the Natural Resources Defense Council.

Cavanagh was a rather patrician New Englander who had been turned on to energy policy by a mentor at Yale Law School. He had joined NRDC as a young attorney in 1979 and had never left. (He was in the San Francisco office the day the pro-nuclear marchers sat down to block the building's front doors, and he departed through a different exit.) He saw himself as a pragmatist who, when it came to nuclear power, looked at the specific circumstances to determine what made sense. NRDC was generally critical of nuclear, but they had endorsed, for instance, legislation that included support for the continued operations of nuclear plants in Illinois. Cavanagh also, in contrast to some environmentalists, saw utilities as potential partners rather than automatically as adversaries. He had a long-standing relationship with PG&E, which was why Freeman wanted him involved.

Cavanagh agreed that in California, phasing out nuclear was the way to go—that because of the state's progress on renewables, it would be possible to shut down Diablo Canyon without increasing fossil fuel emissions. Starting in 2002, the state had passed a series of landmark laws requiring utilities to meet certain renewables targets. Most recently, in 2015, the state legislature passed Senate Bill 350, which mandated that utilities procure 50 percent of their electricity from renewable sources by 2030. PG&E had already made impressive gains. From 2007 to 2010, for example, PG&E signed contracts for the construction of almost three gigawatts of solar power. At the time, the entire rest of the United States had only half a gigawatt of solar, and the whole world had a total of eight gigawatts.[14] By 2011, the state had a gigawatt of solar power—enough to power 750,000 homes, prompting some to call California "the Saudi Arabia of sun" (although, of course, the panels only provided energy when the sun was shining).[15]

Meanwhile, the growth of renewables was not necessarily compatible with Diablo Canyon's continued operation. The plant provided a large chunk of steady, always-on electricity, known as "baseload power." (A caveat is in order here, too, since nuclear plants have regular and sometimes unplanned outages, but as we've seen, they have the highest capacity factor of any energy source—more than 90 percent in recent years.[16]) This offered obvious advantages, but as solar generation soared, PG&E would often end up with more electricity than it could use during daylight hours, and unlike natural gas, the supply of electricity from nuclear could not be easily ramped up and down to accommodate changes in the solar output. The upshot was that sometimes solar and wind producers had to curtail generation.[17] "There are more and more hours when there is an excess of zero-carbon energy," Cavanagh said. The electricity from Diablo Canyon was getting in the way of taking advantage of the new energy from renewables.

The grid of the future, Cavanagh believed, would be flexible, distributed, and integrated. In this vision, when solar began to decline in the evening, California could rely on battery storage, but also bring in electricity from windy states, or hydropower from Canada, through a grid that would be interconnected over a large region. The state could also deploy "demand response" to financially incentivize individuals and businesses to use electricity at times when it was more available, and to automate some of this use through "smart" technology.

"These giant twenty-four-seven power plants are increasingly obsolete," Cavanagh said. He also noted Diablo Canyon's retro technology: "If you go to the control room, it's an analog control room. It predates the digital revolution." In other words, the American nuclear fleet was starting to look prehistoric.

A series of meetings began with PG&E representatives at the utility's office building, a skyscraper in the heart of San Francisco's Financial District. They met in a conference room on the 32nd floor with marvelous views of the city. Since Freeman was based in Washington, D.C., he usually attended virtually.

Cavanagh and Freeman both insisted that, while they wanted to phase out the plant, they were working to avoid an abrupt shutdown as much as to accelerate the closure. To continue operating, Diablo Canyon also needed new permits from a couple of state agencies, the State Water Resources Control Board and the State Lands Commission. They thought that, by proposing to shut the plant in 9 years instead of 29, they would please some of the anti-nuclear members of those agencies, who would then be more likely to renew the permits for this shorter period.

The meetings went smoothly because, from the outset, everyone was pretty much on the same page about the basic contours of the plan. They held one meeting in D.C., at PG&E's offices there. Someone brought a cake to celebrate Freeman's 90th birthday.

AT THE TIME, PG&E HAD ENTERED A TUMULTUOUS PERIOD THAT CONTINUES to this day. The company had emerged from their first bankruptcy in 2004 (it's never a good sign when you have to qualify "bankruptcy" with an ordinal number), and in 2016, it was in the midst of its first felony trial (ditto). To be fair, the first bankruptcy was not entirely, or even mostly, PG&E's fault: it had resulted directly from the disastrous deregulation experiment that California had undergone in the late 1990s. Deregulation created conditions that were ripe for manipulation by bad actors such as the notorious Houston-based company Enron; it led to skyrocketing electricity costs and rolling blackouts, and culminated in the 2003 recall of Governor Gray Davis. (The felony charges followed a 2010 natural gas pipeline explosion in the town of San Bruno, which killed eight and injured many more. The company was charged with violating the Natural Gas Pipeline Safety Act and would eventually be convicted.)

The utility's stated reason for embracing Friends of the Earth's idea was primarily that Diablo Canyon no longer fit their needs in a changing energy landscape. In addition to the new state renewables requirements, another factor was the growing popularity of Community Choice Aggregation. Made possible by a 2002 state law, these were local authorities that municipalities could establish to procure their own energy. CCAs could set more ambitious renewables targets, and they enabled communities to escape from the monopoly of their default utility. They were deeply appealing to many California communities, promising to realize the vision of local, democratic control and clean energy that environmentalists had long dreamed of. By 2014, there were a number of these in PG&E's service territory—in Marin, Napa, Sonoma, Mendocino, and Solano Counties, among others—which meant that PG&E was losing customers. This trend compounded the dilemma introduced by the renewables increase: if Diablo Canyon stayed online, PG&E might end up with even more excess electricity.

In addition to the environmental groups, PG&E sought to get the Diablo

Canyon workforce on board. For that, they reached out to Tom Dalzell, a veteran labor leader who had worked for Cesar Chavez at the United Farm Workers. Since 1984, he had been with the International Brotherhood of Electrical Workers (IBEW), which represented the majority of Diablo Canyon's employees. One day, he got a call from one of the negotiators at PG&E. He'd already suspected that PG&E was planning to shut Diablo down. While he knew that some of the employees, especially the younger ones, would be disappointed, he told me that rather than engage in a long-shot attempt to influence the decision, his priority was to get a great deal for his workers. This didn't turn out to be hard. The initial offer was amazingly generous—better than he would have asked for. It included retention pay—25 percent of base pay on top of the original salaries—during the period before the closure, severance pay equivalent to a year's salary, and retraining to work on decommissioning or elsewhere at PG&E, at an estimated cost of $350 million. "I didn't play a game," Dalzell recalled. "I said, 'That's good.'"

The deal would need approval from the State Lands Commission, of which then–Lieutenant Governor Gavin Newsom was a member.[18] Newsom had a strong relationship with Dalzell and IBEW, dating back to the 1990s, when Newsom had been on the San Francisco Board of Supervisors. To discuss the proposal, Dalzell met Newsom at one of the restaurants Newsom owned in San Francisco, in the Pacific Heights neighborhood. Dalzell brought his teenage daughter and her best friend, who wanted to meet the celebrity politician. Dalzell wanted Newsom's assurances that he would support the deal; Newsom wanted Dalzell's blessing to do so. Since they were already on the same page, their conversation was brief; Newsom then spent 40 minutes talking with the two teenagers.[19]

Over time, a few other groups joined the talks, including the Alliance for Nuclear Responsibility, which had spun off of Mothers for Peace to focus on the economic case against nuclear. David Weisman of that group recalls going to PG&E's headquarters for a meeting. Outside the conference room were platters of sandwiches. But they were cut into quarters. He was staring at the chicken salad and roast beef and Swiss mini-sandwiches, hesitating to pile them onto his napkin for fear of looking gluttonous. An attorney for another organization, passing by in the hallway, noticed his hesitation. "Yeah, I know what you're thinking," the attorney said, as Weisman recalls. "Go ahead. Take it. It's the last free sandwich you're ever getting from PG&E."

Within six months, the parties had ironed out the details. They agreed that instead of renewing the reactors' NRC licenses for another 20 years, PG&E would close both units when their current licenses expired, in 2024 and 2025. This time frame would afford PG&E and the state almost a decade to replace the electricity with renewable sources and storage systems while also seeking opportunities to increase efficiency. Given various uncertainties about how energy options would develop in the years to come, the agreement (known as the Joint Proposal) stated that "the Parties cannot, and it would be a mistake to try to, specify all the necessary replacement procurement now[.]" That is, the intention was to avoid any new fossil fuel emissions, but the details were still to be determined—and the agreement was not legally binding.

The parties saw the Joint Proposal as a major victory for workers and the climate. They also saw it as a blueprint for how other communities around the country and the world could phase out their nuclear plants in a thoughtful, responsible way. "The Diablo Canyon agreement should put an end to the debate about nuclear power," Freeman wrote in an op-ed in the *Sacramento Bee*. "It provides a template for a timely transition for all the nuclear and fossil fueled plants to an all-renewable future."[20]

As ever, Diablo Canyon was a symbol. It wasn't easy for Friends of the Earth to endorse keeping the plant open for almost another decade, and it wasn't easy for the unions to support a plan to close it at least 20 years sooner than they'd expected. But on both sides, they were able to put aside their qualms. Cavanagh marveled, "Most of the people who'd been fighting each other for forty years over whether the plant should be there, they got together in the end around this solution." A new era appeared to be dawning in California and, maybe, the world.

CHAPTER TEN

"THE MOTHERLY SIDE OF NUCLEAR"

WHEN HEATHER HOFF FOUND OUT WHO WAS BEHIND THE SAVE DIABLO Canyon website, she and Kristin already knew who Shellenberger was; they had seen *Pandora's Promise*. Learning about him and the handful of other pro-nuclear advocates at the time—seeing their ideas validated in such a high-profile way—had reinforced their nascent belief that what they were doing was good and important. Kristin had also seen Shellenberger debate Ralph Nader on *Crossfire*. She was traveling for work; she had just checked into her hotel room and turned on the television, and she sat down, riveted. As she later put it to me, "When are you ever on the edge of your seat during CNN's *Crossfire*?" She thought, *Ralph Nader, you think you know about climate, but this guy just schooled you.*

Heather, although intimidated by the prospect, decided to get in touch with Shellenberger. On February 10, 2016, she sent him an email. "I would love to help, however possible," she wrote. "I keep telling leaders here at the plant that we need to give employees some reasons to be excited rather than only talking about the challenges. The consistent message for how we can help with license renewal is basically—keep doing what you're doing, and don't mess up. That's not very inspiring."

Shellenberger replied right away and forwarded an invitation to a talk for plant employees that he was planning. "As I have come to understand why nuclear energy is the most important environmental technology of the 21st Century," he wrote, "I have come to feel grateful to nuclear power plant workers for your service to society and the environment. I feel your work is not only unappreciated but disrespected by too many people, including by my former self." He went on, "I believe that the people of California, the Governor, and the CEO of PG&E will do the right thing once they consider the large social, economic and environmental benefits of keeping Diablo Canyon running. And that will require a coalition effort not just by environmentalists but also by workers." Heather circulated the invitation among her colleagues and baked about two hundred chocolate chip cookies on the day of the event.

On the evening of February 18, 2016, a couple hundred attendees filed into a conference room at a local Courtyard by Marriott hotel. An image of Rosie the Riveter was projected onto a large screen, with her clenched fist, red polka-dotted bandana, and speech bubble proclaiming, "We Can Do It!" Heather stood near the entrance, greeting people as they arrived. Most of them were plant workers, but one man in a suit, Heather learned, was the superintendent of San Luis Obispo Coastal Unified School District. He wanted to keep abreast of the situation because the schools stood to lose millions in funding from PG&E taxes if Diablo closed; the utility contributed about $22 million in annual property taxes to the San Luis Obispo community. Shellenberger told the audience that Diablo Canyon was essential to meeting California's climate goals, and that it could operate safely for at least another 20 years. He said that it was at risk of being closed for political reasons, and urged the workers to get fired up and organize to save their plant.

After Shellenberger's talk, Heather and Kristin lingered, folding up chairs and ruminating.

"Kristin, do you want to do this?" Heather recalls asking her.

"Yeah, we should do this."

"Yeah, I think we should do it, too. . . . We're the ones that are going to make this happen, we're going to do it together."

Initially using the name Save Diablo Canyon, they organized a series of meetings at a local pipe fitters' union hall. They served pizza for dozens of employees and their family members, who wrote letters to the State Lands Commission and other California officials. Part of their motive, of course, was that they didn't want to lose their jobs. But, Kristin said, "If it was just about our jobs, I would have started polishing my résumé." They believed that, in a state that claimed to be a climate leader, prematurely shutting down Diablo Canyon—which provided almost a quarter of California's low-carbon energy at the time—was absurd.

To brainstorm and plan, the women would meet at each other's homes and at cafés, often with their children in tow. Before long, they decided that their mission was bigger than rescuing their own plant. They wanted to correct what they saw as false impressions about nuclear power—impressions that they had once had themselves—and to try to shift public opinion, to show that nuclear energy actually aligned with environmentalist goals.

Kristin and Heather also knew they brought certain advantages to the effort. As women, they might be able to reach people who would never listen to the bad boys. With Shellenberger's encouragement, they decided to form their own nonprofit. Like the leaders of many other movements led by women—protests against war, drunk driving, and, of course, *against* nuclear power—they sought to capitalize on their status as mothers. They toyed with a few generic names—Mothers for Climate, Mothers for Sustainability—because they worried that the word "nuclear" would scare some people off. But they ultimately discarded those more innocuous options. They chose a name that sounded like they'd gotten the preposition wrong: Mothers for Nuclear. They officially launched on Earth Day of 2016.

They wanted to turn all of the tropes about women and nuclear on their heads. As they saw it, nuclear energy was a life-giving force. They emphasized the value of reliable electricity—and here they did depart from the standard environmentalist line, which typically stressed conservation of energy, not the benefits of its abundance. For hospitals, for people who needed oxygen machines in their homes, even for those who needed air-conditioning during increasingly frequent heat waves, lack of electricity could kill. And reliability was an advantage that nuclear had over solar and wind, which, in the absence of widespread storage, were dependent on certain weather conditions to operate. For stable round-the-clock electricity, at least with the available technology, the options were nuclear and fossil fuels—and fossil fuels killed, too, more than most people realized. Even leaving aside the massive harms of climate change, particulate matter from the burning of fossil fuels was blamed for up to eight million premature deaths per year globally.[1] By contrast, estimates of the deaths attributed to Chernobyl, the worst nuclear accident in history, ranged from the dozens to the thousands.[2] As far as Kristin and Heather were concerned, the biggest problem with nuclear was its public image. And they were determined to be the ones to change that image.

Shellenberger, with his background in PR, sensed their advantages as spokespeople, too. True, conventional wisdom suggested that their industry connection would be a liability, enabling people to write them off as shills just out to save their jobs. But Shellenberger thought, as he said in a podcast interview, "If the workers aren't advocating for their plants, you start to have questions about how good the plants are." He was also acutely aware of the gender dynamics related to nuclear. When he first met Kristin and Heather, the night of his talk at the Courtyard by Marriott, he was struck by "these two beautiful young mothers" who were passionate about nuclear power. Who could be better ambassadors, better faces of nuclear?[3]

To get their message across, they set up a website featuring photos of children and of scenic outdoor landscapes, including the area around the plant. It was an effort to associate nuclear energy with clean air and water, conservation and open spaces. Cognizant that much of the pro-nuclear rhetoric was dominated by men who could be aggressive and condescending, they emphasized that they wanted to engage in dialogue: "We used to be skeptical of nuclear power," the home page read.

They thought the nuclear industry had done a terrible job of communicating with the public. For one thing, it had emphasized safety, which only served to activate fears; airlines, they pointed out, don't advertise by touting their safety records. Kristin and Heather wanted, instead, to unapologetically celebrate nuclear for its strengths. They also disliked the tendency of industry representatives to respond to criticism with a barrage of information and numbers and charts. "I think about this a lot as a mom," said Kristin. "I want to know what is dangerous for my kids," but "sorting through the data is so hard." She wanted to be a trustworthy messenger who could just tell mothers and others what they needed to know. "We do a real disservice to people when we overblow risks." As for nuclear, she said, "Yeah, there are risks, but they're relatively minor. We get people so focused on things that aren't actually the biggest threats to them and then they end up underplaying or ignoring the things that actually are the biggest threats to them."

On their website, they each had a page telling their stories: about their families, their jobs, why they changed their minds about nuclear. They had photos of themselves: long hair, makeup-free faces, bright smiles. As outdoorsy, almost hippie-ish moms, they were not your stereotype of a nuclear worker, or, for that matter, a nuclear supporter. They certainly didn't offer a barrage of data; the website contained surprisingly little in the way of hard facts, and sometimes their desire to be clear and unapologetic about the benefits of nuclear led to dubious statements. "Nuclear power is the safest way to generate grid electricity, period," the FAQ page read. "There is zero debate among public health researchers and scientists about this question."[4] There were no links to studies backing up these claims, and the assertion of "zero debate" was wishful thinking at best.[5]

The two women aimed to represent "the motherly side of nuclear," as Heather once put it. Early on, they enlisted a designer to create a logo. Like the Mothers for Peace poster, their logo shows the outline of a mother holding a child—but the two figures are framed by the symbol for the atom.

THE DAY BEFORE THE PROPOSAL TO CLOSE DIABLO CANYON WAS announced, Heather and Kristin were in Costco, stocking up on coffee and oatmeal, hot dogs and buns. They were preparing for the pro-nuclear march, which was scheduled to begin in four days. They knew that the plant was in trouble—that was the main motive for the march—but no specific plan had yet been announced. In the aisle at Costco, Kristin got a text message from a friend, alerting her that PG&E was going to make a big announcement the next day.

At 6:00 the next morning, June 21, 2016, an email was sent to all plant employees announcing the Joint Proposal, the agreement reached with labor and environmental groups to close Diablo Canyon in 2025. At the plant, PG&E leaders were on-site from 5:00 a.m. to 9:00 p.m., meeting with employees to explain the plan and answer questions. The mood was somber.

Many people who had planned to join the march canceled. Kristin and Heather wondered if they should call it off. But not only had they already spent hundreds of dollars of their own money at Costco, they had borrowed gear for camping on the way from San Francisco to Sacramento. "We had planned, we had spent money, and we had people coming from out of state to help us," said Heather. They thought, *Yeah, we still have to do this.*

On the morning of Tuesday, June 28, 2016, Heather and Kristin and some of the other marchers, the ones who had made it all the way to the end of the odyssey, arrived at the Holiday Inn Capitol Plaza in Sacramento. They had camped for four nights at a county park, Lake Solano, holding strategic meetings and marches at nearby locations during the day. Here in Sacramento, the State Lands Commission would hold one of its regular, bimonthly public meetings. At this one, they would be deciding whether to extend a lease needed to continue operations of Diablo Canyon, and for how long. PG&E was requesting a new lease that would be coterminous with its NRC licenses expiring in 2024/2025. In short, this meeting was one key step in the process of navigating the state's extensive environmental oversight apparatus in order to carry out the Joint Proposal that the parties had agreed to.

The meeting was scheduled to begin at 10:00 a.m., in a drab room with patterned beige wallpaper. At 10:06, State Controller Betty T. Yee, the chair of the commission, called the meeting to order, sitting at a table with a black-and-green paisley table skirt, facing the members of the public who had filled the maroon chairs. It was the kind of unglamorous, almost kitschy setting where these interminable hearings often take place. When Newsom arrived late, with his perfectly coiffed hair and slim-fitting suit, he looked out of place; this seemed like a particularly dreary but necessary stop on his glide to higher office.[6]

First came the warm and fuzzy and mutually congratulatory portion of the meeting. Geisha Williams, president of PG&E's electric division, wearing black heels, pearls, and glasses, stood at the lectern and thanked the commission "for really challenging us to think about a different type of clean energy future for California. And I believe that what we're proposing today really delivers on that challenge." She referred multiple times to the "certainty" that approving this plan would provide for the plant's employees.

Representatives from the parties who had signed on to the proposal came to the lectern to voice their endorsements—"I mean, it is historic that Friends of the Earth and PG&E and Local 1245 are all here agreeing to something," said Tom Dalzell.

And then it was time for public comments. Although, as Cavanagh and Dalzell and the other supporters noted, a number of adversaries had come together behind the deal, not everyone had. Many of those patiently sitting in the uncomfortable chairs—including members of Mothers for Peace and Mothers for Nuclear—opposed the plan, but for opposite reasons. The pro-nuclear people wanted Diablo Canyon to stay open much longer, while the anti-nuclear people wanted it to close sooner—or at least be subject to an environmental impact review, from which PG&E was requesting an exemption.

"What could possibly be wrong with a fact-based, scientific decision?" asked a white-haired man named Glenn Griffith, who said he was associated with Mothers for Peace and the anti-nuclear World Business Academy. "I am so tired of all this politics. I am so tired of all this cancer. My wife has breast cancer. She's going through chemo. She's going through radiation, mastectomy. I'm so tired. I'm so tired of standing before all these commissions asking, begging on bended knee for you people to do the right thing."

When people testified, it wasn't hard to tell which side they were on. The anti side brought up Fukushima, nuclear waste, aging components, terrorism, and PG&E's untrustworthiness ("PG&E is organized crime."). The pro side mentioned baseload power, grid reliability, the intermittency of renewables, fearmongering, jobs, and tax revenue. The anti side cited Germany—which was phasing out its nuclear plants and transitioning to renewables—as a model. The pro side—contending that this transition was not going very smoothly—denounced Germany as a disaster.

And so it went. A number of people mentioned earthquake risk; when it was Kristin's turn to comment, she said, "As a professional civil engineer, I can personally testify to the plant's robust seismic design. I feel comfortable working there and having my family live nearby. I cannot say the same thing for the rest of California."

When Heather went to the lectern, her six-year-old daughter, a blond sprite, climbed into her arms, and Heather held her as she spoke. "I support renewables, but I believe that every technology has a place," she said. "Let's keep Diablo Canyon's power and use renewables along with storage to replace fossil fuels instead."

Both sides seemed to think their testimony was futile, that this was a backroom deal awaiting a rubber stamp, that the whole hearing was a sham. "Any questions?" asked one older man, who identified himself as a geologist, after detailing the seismic dangers. When Yee quickly said no, he replied, in a voice dripping with contempt, "I didn't think so."

After almost four hours, Newsom addressed the audience. Sounding a bit defensive, he noted that the deal had been well received nationally: "I mean, Bernie Sanders himself tweeted, 'This is a model for the nation.'" Newsom acknowledged that the closure of SONGS was "a disaster. It increased wholesale energy costs, it hurt working folks, it hurt the economy, and it increased greenhouse gas emissions." He insisted that he wasn't ideologically opposed to nuclear power. He said he was "good friends with"—wait for it—Stewart Brand, who had regaled him on the merits of nuclear at his houseboat in Sausalito.

But, Newsom went on, this plan was farsighted and reasonable. "And I think this will provide a bridge where San Onofre was a ditch," he said. "And we have the opportunity to hold folks accountable and do something I think that, at the end of the day, we can all be proud of." He moved to approve the item, and the other two commissioners agreed. "Without objection," said Yee, "such will be the order."

CHAPTER ELEVEN

THE NUCLEARISTS

ALMOST THREE YEARS LATER, IN MAY OF 2019, I FOUND MYSELF STANDING in a dark stairwell, wearing a hard hat and peering into the control room at Diablo Canyon. My guides, Heather Hoff and Kristin Zaitz, were standing next to me. Through a small window in the door, I could see a large avocado-green panel, studded with knobs, buttons, and switches. A man was standing at the panel, and a few others were sitting around in chairs. They had what seemed to me an unenviable responsibility: monitoring hundreds of signals to ensure that the entire plant was running smoothly and that any glitches were promptly addressed. "A boring day is a good day," said Heather, who had spent countless hours in that room.

By this time, the plan to close the plant had been approved by various agencies and the legislature. As for me, I had recently begun to look into the question of nuclear power in earnest. It hadn't taken me long to identify the small but vocal group of people who were touting its virtues—an enthusiasm I was still trying to wrap my head around. When I came across Mothers for Nuclear, the name, as it was intended to do, caught my attention.

On their website, I saw photos of Kristin backpacking in the Sierras with her kids, of Heather canoeing in a lake with her dad; I read their personal stories. Clearly, I was their target audience: a woman, a mom, who was worried about climate change and had vaguely bad associations with nuclear but was not staunchly opposed to it. In some ways, I felt a kinship with them. Heather wrote about her visceral aversion to wastefulness and her obsession with recycling—propensities I shared but had learned that most people found annoying. The two women seemed like more outdoorsy and capable versions of me, the kind of person I sort of wished I were. They seemed to bring to camping and hiking the same competence that (I hoped and presumed) they applied to their jobs. You wouldn't want to be lost in the woods with me, just as you wouldn't want me presiding over the control room at a nuclear plant, but I got the sense that, with them, you might make it out of both scenarios alive.

Of course, I knew to be cautious about accepting their claims at face value. I recognized the patent self-interest in their desire to change public opinion and to rescue the plant that paid their salaries. Every so often, I was gripped by a jolt of suspicion. *Wait a minute. Is this a trick? Mothers for Nuclear? Come on!* But I was intrigued enough to want to learn more. I figured their jobs must give them, if not an unbiased understanding of nuclear, at least a particularly intimate one. And they could offer me something that most nuclear advocates could not: a tour of a nuclear plant. Could a trip to Diablo Canyon, and a day with Mothers for Nuclear, win me over?

Back in January, I'd spoken with Eric Meyer, the affable millennial who had organized the March for Environmental Hope. He had since moved from the Bay Area back to his home state of Minnesota, where he'd founded a small nonprofit called Generation Atomic. The website featured quotes from prominent nuclear supporters such as Steven Chu, President Obama's Secretary of Energy ("If we want to make a serious dent in carbon dioxide emissions . . . then nuclear power has to be on the table") and Cory Booker ("Nuclear energy should be part of the decarbonization toolbox"). The group's Twitter feed was full of witty pro-nuclear graphics and memes. I watched a video of Eric singing at a rally in Amsterdam, in his deep baritone, "I can't help falling in love with you." On the last note, he made his right hand into the shape of a U, to indicate that he meant the element (uranium), not the pronoun.

Clearly, Eric was trying to bring to nuclear power something for which it had never exactly been known: a sense of fun. As the name of his organization implied, he believed that anti-nuclear sentiment was a relic, an attitude for clueless Boomers, for aging hippies. (Strikingly, I noticed that similar insinuations went the other way; anti-nuclear advocates saw atomic energy as a technology of the 20th century, disdaining support for it as backward-looking.) Eric had faith that the younger generation—without the Cold War baggage, and with an appropriate degree of alarm about climate change—would be open-minded and unbiased. They would look at the evidence about nuclear and, like he had, fall in love.

Eric grew up in a small town in southwestern Minnesota near Buffalo Ridge, a plateau of high-altitude prairie known for its strong winds. Even back then, in the '90s, spinning turbines punctuated the landscape. He assumed they represented the future. Meanwhile, as with Kristin Zaitz, his only impressions of nuclear energy came from the leaky, shambolic plant on *The Simpsons*. In college, though, he learned how far we were from meeting our clean-energy needs: the vast majority of the world's energy was still derived from fossil fuels. He also learned that much of humanity still lacked the basic electricity that Americans took for granted. In the face of the need, the wind turbines on the prairie started to seem a little puny. In 2010, a friend sent him a link to a TED Talk about a next-generation concept—a molten salt reactor—saying, "Hey, check this out, we might be able to power the future on nuclear waste and dismantled nuclear weapons." Before long, Eric was a full-fledged convert, determined to save the world with nuclear energy.

Plenty of people are torn between career paths: public-interest attorney or corporate lawyer? Plumber or electrician? Eric was possibly the first and only person to agonize over whether to become a nuclear power advocate or professional opera singer.[1] For a while, he tried to juggle both ambitions, applying to two masters programs, one for vocal performance and another for advocacy and political leadership. But he soon realized he could only afford to pay for one of them. At the same time, he had been cast in a production of *La Traviata*, but some of the rehearsals conflicted with a nuclear conference he wanted to attend. Ultimately, he concluded that the world needed nuclear power advocates more urgently than it needed another opera singer.

So in 2016, at age 28, he drove his used Ford C-Max hybrid hatchback out to the Bay Area. He found an overpriced studio apartment on Telegraph Avenue, in the heart of Berkeley, and started working for Shellenberger at Environmental Progress. The next year, he left to cofound Generation Atomic with a friend. Shellenberger had begun to work on a book and wanted his employees to help with research, while Eric wanted to continue focusing on grassroots organizing. The parting was amicable; Shellenberger wrote him a check to help him get started.[2]

Like Ted Nordhaus, Eric had experience with canvassing. In his case it had come from his volunteer work in 2012 urging voters to reject a constitutional amendment to ban same-sex marriage in Minnesota. His main motive was to support his dad, who had come out as gay when Eric was eight years old. In small-town Minnesota in the 1990s, this was not a shrug-worthy turn of events. His dad was the first openly gay man in the town. Eric endured cruel mockery from classmates and even a visit from a pair of Child Protective Services workers, who interrogated him about whether his father had abused him. By 2012, his dad had a longtime partner, and they wanted the right to marry. Eric talked to hundreds of voters, often invoking his dad's experience. In the end, the amendment was defeated.

In Ohio, Eric applied those skills to saving two nuclear plants, Perry and Davis-Besse, that were struggling economically. He and his cofounder, Taylor Stevenson, stayed in sketchy motels night after night, sharing beds to save on costs. They designed and printed postcards outlining reasons to support the plants: "Nuclear energy provides 4,300 community sustaining jobs"; "Provides More Than $23 million in direct funding to Schools"; "Generates 90% of Ohio's Clean Energy." When they knocked on doors, they asked the residents which reason mattered most to them, and urged them to write a message on the postcards to their elected officials. Eric persuaded FirstEnergy Corporation, the owner of the plants, to provide some funding, enabling them to hire about 20 canvassers. Altogether, from late April through September of 2017, Eric says they knocked on more than 44,000 doors, engaged in roughly 8,700 conversations, and mailed 8,465 postcards. By 2019, the plants were still operating but their status was in limbo.

When we first spoke, early in 2019, Eric alerted me to an upcoming conference on advanced reactors at UC San Diego, not far from my home. I drove there on a rainy day in January. The conference was held in the vast student center on campus, in a room with a stage and a looming screen for PowerPoint slides. The presentations, with titles like "U.S. Nuclear Regulatory Commission Transformation & Licensing Modernization" and "From Very Small to Mega: The Right Reactor Size for the Right Application," were highly technical. I stayed mainly in the area outside this room, where people were mingling around tables covered by black tablecloths, with pamphlets on display. Most of the exhibitors were advanced nuclear start-ups (NuScale Power, Terrestrial Energy) and government entities, such as the Oak Ridge and Idaho National Laboratories.

There, I met Eric in person, by a table with Generation Atomic pamphlets ("Engage→Train→Empower") and merch. He had black-framed glasses and a mustache (I couldn't tell if it was ironic). He introduced me to some of the others. I met Mary Anne Cummings, a particle physicist with wild gray hair who had served as a Bernie Sanders delegate at the Democratic National Convention. (She told me that nuclear was the one issue where Bernie got it wrong.) I met Eddie Guerra, a Puerto Rican–born engineer. He was working to ensure that small modular reactors would play a role in rebuilding the island's energy infrastructure, which had recently been devastated by Hurricane Maria. We chatted with Dr. Rita Baranwal, director of the Gateway for Accelerated Innovation in Nuclear (GAIN) initiative at Idaho National Laboratory, who would soon be appointed to a leading role at the Department of Energy, and Jackie Kempfer, a young woman who worked at Third Way. "It's like magic!" she said of nuclear energy. "This is like Harry Potter, people!"[3] Pretty much everyone seemed to know one another. Evidently, in the small world of nuclear advocacy, attending a nuclear conference was a bit like a regular on *Cheers* showing up at the bar.

After the conference, Eric wanted to visit SONGS, the plant not far from San Diego that, he said, had closed too soon; he wanted to pay his respects. He was without a car, and I offered him a lift. By late afternoon, the rain had cleared, and the air had the freshness that comes only after a storm. We parked in the lot at the state park adjacent to the shuttered plant and began walking on the pristine beach, stepping over puddles in the sand. Virtually no one else was there. Thick rays of sunlight broke through the heavy clouds above the horizon. As we walked, Eric sang an adaptation of a song from *Faust*, with lyrics he had written from the first-person perspective of fossil fuels. (Keep in mind that when he sings, he sounds more or less like Pavarotti.)

> *"When you heat your home, you burn me up*
> *When you drive your car, up I go in flames. . . .*
> *The wind and sun are so alluring, but alone won't do the trick*
> *Sun goes down, wind stops blowing*
> *And when they do, I'll make you sick."*

After about 20 minutes, the massive concrete structures of the plant came into view. We observed a moment of silence.

When we returned to the parking lot, we saw a man wearing a border patrol uniform and exchanged greetings with him. Eric told him why we'd come, explaining that he was an environmentalist, so he wished the nuclear plant had remained open. The guy looked at him. "Is that a joke?" he said.

By this point, I was beginning to understand the case for nuclear power. The pluses were clear: the small land footprint, the independence from weather conditions, and of course, the lack of greenhouse gas emissions from its generation. I was still uncertain, though, how serious the downsides were. To be sure, it seemed possible that the older generation of anti-nuclear activists had (understandably) overestimated the risks of this technology when it was unfamiliar and chiefly associated with apocalyptic weapons. But the opposite response also seemed plausible: that now, with the memories of the Cold War receding, and growing desperation to address climate change, a younger cohort would grasp onto a solution that seemed promising and wave away the perils. (Surprisingly—to me, anyway—survey data showed that younger people on average had more negative impressions of nuclear than older respondents,[4] but I found that most of the pro-nuclear activists were on the younger side.)

I was still pondering these questions in May, when I drove up the coast to San Luis Obispo. Since the '60s, its population has grown to over 45,000. Its downtown is still centered around Mission Plaza, which features a large adobe church, built in the late 1700s, with a silver cross rising from the ridge of the red-tiled roof. A creek burbles nearby. The city's street signs are charmingly idiosyncratic, with the names—Peach, Chorro, Monterey, Higuera—written in a distinctive white Gaelic font against a green background. Local breweries and cafés mingle with upmarket chains like Lululemon and Williams Sonoma. A couple of miles from downtown is the campus of Cal Poly. The city has considerable charm, as does the broader Central Coast region, with its beaches and hiking trails and wineries; it offers a slower pace of life, and more affordable real estate, than the major cities to the north and south. I would later be told that this location was a prime reason many people wanted to work at Diablo Canyon.

In advance of my visit, I had submitted information for a background check, and Heather Hoff had instructed me to bring my passport and to wear closed-toe shoes. Was I nervous about visiting a nuclear plant? Oddly, not really. Historically, I was not an unanxious person, and, although I never had a particularly acute fear of nuclear power, I did go through a phase of preoccupation with toxins of all kinds. This was in 2012, after my daughter was born, and my ears were suddenly alert to rumors about the endocrine-disrupting chemicals in plastics, in cookware, even in kids' pajamas. When I looked around me, all of the material objects I saw seemed full of menace. At that time, I would have been consumed by dread at the prospect of visiting a nuclear plant, sure that I would somehow become radioactive and contaminate my daughter. Since then, however, I had effectively maxed out on that neurosis—I'd become resigned to the fact that toxins were everywhere but had managed to stop worrying about it most of the time. What was more, I had been talking with pro-nuclear advocates for several months already, and listening to their assurances surely predisposed me to feel sanguine.

The morning after my arrival, I woke up at my hotel, got ready, put on my closed-toe shoes, and went down to the lobby. Outside, the day was drizzly. Heather came through the door, smiling and holding a travel mug. We went out to her car, a slate-gray electric Ford Focus, adorned with a "Split Don't Emit" bumper sticker, and drove to a nearby café.

Over coffee and an almond croissant, Heather told me about her father, a glass blower and "mad scientist" type. When she was in elementary school, "the class would go on field trips to *my house*," she said—her trailer in the Arizona desert. He would show the children his collection of gems and fossils and minerals, beakers and test tubes. He would breathe in helium and entertain the kids by talking in a high-pitched voice. Notwithstanding this playful streak, he'd been pessimistic about the future of humanity and the planet. He had discouraged Heather from having children, because he thought life would be hard for the next generation. But after Heather started working at Diablo Canyon, she transmitted to him her excitement about what she was learning. Before he died of a rare form of dementia at 71, he had become curious about nuclear, Heather said, and his fatalism had softened.

As we waited for Kristin to arrive, Heather showed me the pendant she was wearing on a chain around her neck. Crafted for her by an artist her dad had mentored in Arizona, it was made partly of uranium glass, an old-fashioned material that has a touch of uranium added in for aesthetic purposes—it gives the glass a faint purplish cast. It was a great conversation starter, she told me: "I wear it as a demonstration—radiation is not necessarily dangerous. It's all around us." This, I would learn, was a key theme for nuclear advocates. They would point out that you would get 10 times more radiation from one CT scan than nuclear workers typically get in a year;[5] they'd hand out bananas at rallies, explaining that they contained potassium-40, a mildly radioactive isotope.

When Kristin appeared, Heather brightened and handed her the coffee she'd bought for her. Since starting to work together on their nonprofit, they had become close friends who clearly enjoyed each other's company; they reminded me of podcast cohosts who have really good chemistry, riffing and making each other laugh. When we set out for Diablo Canyon, I rode up front; Kristin sat in the back, pumping breast milk for her year-old daughter. The light rain had stopped, but mist still hung in the air. We passed through the town of Avila Beach, driving alongside the ocean. To our left, aquamarine water sparkled. To our right lay gently sloping terrain of grasses, sagebrush, wildflowers, and shrubs. Along the curving road, a sign proclaimed, "Safety Is No Accident." In the distance, the two gargantuan containment domes rose above a cluster of shorter structures.

After stopping at the security checkpoint and continuing along the road, we at last pulled into the parking lot. We walked over to one of the outbuildings, where I handed my passport to a blond woman sitting behind a counter. Entering a nuclear plant is a lot like boarding an airplane, only a lot less crowded. I placed my jacket and my shoulder bag into a gray plastic bin that subsequently made its way on a conveyer belt through an X-ray machine. I passed through a metal detector. Then I stood under the arch of a "puffer machine," which blasted me with air, dislodging loose particles and analyzing them for traces of explosives.

The parallels between nuclear energy and aviation do not end there. It's a commonly invoked analogy but a useful one. Achieving liftoff and splitting the atom were both astonishing scientific feats that seemed to illustrate humanity's growing dominance over nature in the 20th century. And, perhaps in part because of this unnaturalness, both tend to inspire instinctive fear.

At least at first, this fear was justified. Early plane rides frequently ended in fiery crashes. Nuclear fission first took the form of the bomb, and then the young nuclear power industry was something of a make-it-up-as-we-go-along affair, as the Three Mile Island government report underscored. By

2019, however, the safety records for both the aviation and nuclear industries had improved dramatically. As the adage goes, you're more likely to die on the ride to the airport than on the plane. In recent years, roughly 40,000 Americans have died annually in car crashes,[6] but since 2010, there have been a total of two deaths related to accidents on U.S. airlines.[7] Similarly, pollution from coal, oil, and natural gas is estimated to cut short millions of lives per year, while annual deaths attributed to normal operations of nuclear plants hover around zero. (There is some debate about the number of deaths from routine operations of nuclear plants, since workers are exposed to elevated doses of radiation, and even the public can be exposed to small amounts. Some experts believe that these exposures result in some additional deaths, though I haven't found a widely accepted estimate of the number. Interestingly, the same could be said for aviation, because when we fly, we are exposed to elevated levels of "cosmic" ionizing radiation.[8] Still, it seems fair to say that the numbers are relatively low.)

In both cases, these low mortality figures have a lot to do with the much-touted "safety culture" that both industries have built up over time, directly in response to the inherent risk. One way of looking at it, then, is that they're safe because they're dangerous. The flip side is that when something does go wrong—a plane crash, a nuclear accident—it's high-profile and spectacular.

Once it had been determined that I wasn't packing any guns or bombs, we walked to the office building where Heather had a cubicle, climbing the stairs to the third floor. Signs instructed us to hold on to the railing, which we all dutifully did. "Because if someone needs a Band-Aid at a nuclear plant, it's a big story," Heather quipped.

At Heather's cubicle, we put on hard hats and safety glasses, which we needed to enter the adjacent turbine hall. Nearby, we opened the door to "the bridge," a narrow corridor with large windows that connects the administration building to the turbine hall. It was lined with small black lockers for workers to store their own hard hats and safety glasses. Resting on one set of lockers was a plastic bin full of orange earplugs. Heather reached into it and handed me a pair. Through the windows, we could see the ocean, where water was continuously cycling into and out of the plant. A security guard, armed with a handgun and a rifle and wearing a red backpack, sauntered by.

The turbine hall, a vast space with a soaring, arched ceiling, was dominated by two large generators. Outside, within the two containment domes, uranium atoms were splitting apart in a chain reaction, heating water to more than 600 degrees Fahrenheit; the steam spun the turbines, which in turn drove the generators. The resulting electricity would bring power to about three million Californians. Warm air rushed noisily around us. Through the din, Hoff explained different parts of the system: the pipes; the springs that supported them; the condenser, which takes wet vapor from the turbine exhaust and turns it back into liquid. Vending machines selling Pepsi and Chex Mix stood against one wall.

Roaming the space, I was struck by how antithetically different people saw the plant I was standing in. To Mothers for Peace and other local anti-nuclear activists, it was a source of dread, a behemoth lurking nearby, in whose shadow they'd been forced to live; it was another Fukushima waiting to happen. To the plant employees, and to pro-nuclear climate activists, it was a force for good: a source of stable employment and a low-carbon provider of the steady electricity we depend on for our modern lives. I understood both perspectives, and indeed, I realized, to some extent they were not mutually exclusive.

We exited the space, and Heather and Kristin were careful to close the door behind us, explaining that it was a fire door, which would thwart the spread of any blaze from one area to the next. Now we stood in the dim stairwell between the turbine hall and the control room. I was not allowed to enter but I could peek into the small window in the door. Again I was reminded of aviation: the panels lining the walls were like much larger versions of those used by pilots. For several years, after working as a plant operator, Heather worked there as a control room operator.

"Did you stay in there all day?" I asked.

"Yes," Heather said.

"Are there any windows?" I asked.

"No."

"You're asking all the questions I asked before I decided *not* to work there," Kristin chimed in.

Heather was in the control room on March 11, 2011, the day of the tsunami in Japan. She was summoned in to work early; they were on the lookout for tsunami waves that might reach the Pacific coast. Over the next few days, while at the plant, she watched news footage of explosions in Fukushima. What she saw resembled the scenarios she had learned about in training—situations that she had prepared for but never expected to face. "My heart instantly filled with fear," she later wrote on the Mothers for Nuclear website.[9]

For a time, her confidence in nuclear power was shaken. But as more information emerged, she came to believe that the accident was not as cataclysmic as it had initially appeared. "As I learned more about the details of both short- and long-term impacts of the accident," she wrote on the website, "it became clear that our fears were largely misdirected."

Heather and Kristin visited Fukushima in 2018, when they were invited to speak about nuclear advocacy at the Institute of Energy Economics in Japan. Much of what they saw was dystopian: empty towns, damaged buildings, black plastic bags of contaminated soil heaped on the roadside. "Intact reactor buildings next to the twisted metal and crushed concrete of Units 1, 3 and 4," Heather wrote. "Beautiful forests and cherry trees next to wide expanses of asphalt and concrete." She acknowledged, "Thinking of the forces that caused this damage is scary, and it's not something anyone ever wants to be repeated."[10]

For people around the world, the nuclear accident became the event that defined the day, rather than the tsunami itself, which had instantly killed more than 15,000 people and battered much of the country's infrastructure. By 2016, by contrast, the World Health Organization reported that no deaths had resulted directly from the accident at the plant, although they said that about 170 workers might have an increased risk of developing cancer—in most cases, only an imperceptibly slight risk. (In 2018, one worker died of lung cancer.) According to the WHO, "the health risks directly related to radiation exposure are low in Japan and extremely low in neighbouring countries and the rest of the world."[11]

Deaths had followed the accident, but they were nearly all related to the evacuation. More than 160,000 residents fled in the aftermath. Some of these were elderly people hastily relocated from hospitals. Many had to stay for months in unsuitable sites such as crowded public gyms. Ultimately, more than 2,000 people died, according to official reports, as a result of physical and mental stress induced by the evacuation.[12]

One way of looking at these numbers was to lament that the nuclear accident had necessitated such a traumatic and deadly evacuation. In other words, although it was awful to make people leave their homes, the risks of staying would have been greater. But that was not how Heather and Kristin saw it. They believed that staying put would have led to lower risk for the population than leaving—that the problem was the overreaction to the accident more than the accident itself.

This, I was coming to understand, reflected a central tenet of the pro-nuclear worldview. As Eric Meyer told me, "You see time and time again that the fear of radiation, fear of nuclear, has been more dangerous than nuclear itself." The phenomenon was evident, they believed, not only in unwarranted evacuations but in a variety of other ways, most notably the use of fossil fuels instead of nuclear; after Fukushima, for example, both Japan and Germany shuttered some of their nuclear plants and replaced the energy largely with coal.[13]

This view of Fukushima was certainly in the minority, but it wasn't utterly fringe. It found support in a 2017 study by Philip Thomas, a professor of risk management at the University of Bristol. "If no one had been evacuated, the local population's average life expectancy would have fallen by less than three months," wrote Thomas. An average decline of close to three months, in my view, is not a trivial reduction. But as always with nuclear, it's important to keep context in mind. "In another comparison," Thomas went on, "the average inhabitant of London loses 4.5 months of life expectancy because of the city's air pollution. Yet no one has suggested evacuating that city."[14]

Another crucial aspect was stigma: Heather and Kristin spoke with indignant sympathy of people they had met or heard about who were ashamed to admit they were from the area, because others would think they were contaminated; and of fishermen whose livelihoods were destroyed because of the fear that the fish they caught would be radioactive. When the Mothers visited Fukushima, they made a point of eating the local fish—even though Kristin was six months pregnant with her third child at the time.

The argument struck me as worth taking seriously. After all, it's irrefutable that in some cases—the reaction to the terrorist attacks of 9/11, for example—a disproportionate response to a fear or a threat can cause great harm. But there were complications. One was that, in the confusion at the time of the accident, it wasn't clear how severe the radiological impacts would be. After my visit, I interviewed Steve Fetter, who worked in the Office of Science and Technology Policy in President Obama's White House when the meltdowns occurred and was in the room with other decision-makers monitoring the situation. He recalled the "huge amount of uncertainty," especially regarding the state of the spent fuel pools. If they had leaked and the spent fuel had been exposed to the air for an extended period, it could have caught fire and spewed radionuclides. It turned out, in the end, that the water in the spent fuel pools held, and the wind also happened to blow in a lucky direction, carrying most of the radioactive cloud from the reactor meltdowns out to sea. But there was no way to know that at the time. When I mentioned the 2017 study by Philip Thomas, Fetter said, "I think it's the wrong way to look at it. It's of course retrospective. At the time when you had to make the decision whether to evacuate, you didn't know. There was the potential for a much larger release of radiation."

Another complexity involved how to think about the psychological effects of radiation exposure. If you knew or suspected that you'd been exposed to radiation, that awareness could poison your life, even if you were not yet physically sick. A 2015 study in the *Lancet* looked at the aftermath of five nuclear accidents. The authors found that "common issues were not necessarily physical health problems directly attributable to radiation exposure, but rather psychological and social effects."[15]

One might consider the emotional distress an inherent accompaniment to the radiation exposure—part of what makes it distinctively disturbing.[16] But from a pro-nuclear perspective, the emotional distress was, again, a result of excessive fear. After all, didn't the paucity of radiation-related physical health problems found in the study support that view? There's a striking quote in Adam Higginbotham's tome *Midnight in Chernobyl*, an exhaustive account of the 1986 disaster. Years afterward, one of the first responders contemplates the possible effects of the radiation and all of the unknowns. He sums up: "But when my friends ask me about it, I tell them: the less you think about it, the longer you'll live."[17]

As we descended the stairs from the control room and made our way back to the building where we'd entered, Kristin told me that the buildings we had visited were not radiological areas. Still, on our way out, I had to pass through another machine, this time a "radiation portal monitor," to ensure that I was not contaminated. I stood beneath its arches until receiving a verdict from a robotic voice: "Clean."

IN JULY OF 2019, TWO MONTHS AFTER MY VISIT, THE REPUBLICAN-controlled Ohio state legislature passed a bill, HB6, that provided ratepayer-funded subsidies of more than a billion dollars to support two nuclear plants—the ones Eric Meyer had knocked on doors to save—and two coal-fired plants. The bill, which Governor Mike DeWine immediately signed, also weakened existing targets for renewables and energy efficiency. Criticism from the clean-energy world was swift. Leah Stokes, a respected energy expert and a political science professor at UC Santa Barbara, wrote in the *Guardian*, "I have spent the past five years researching states' efforts to roll back clean energy laws." The Ohio law, she wrote, was "the worst yet."[18]

Yet Heather and Kristin had supported the bill, albeit ambivalently. They had retweeted Generation Atomic's entreaty to Ohioans: "CALL YOUR SENATOR TODAY AND ASK THEM TO VOTE YES ON HB6." When I asked Kristin about it, she said that they had not been pleased with the specifics of the legislation. "It's a terrible look for nuclear to partner with coal," she admitted. But they had held their noses because they thought it was so urgent to ensure the continued operation of the nuclear plants, which provided the vast majority of the state's low-carbon energy. "You can't just shut them down and then restart them two years later," Kristin noted. She summed up their feelings: "Ew, but glad they saved the plants."

Over the past half year, I had been surprised again and again by my encounters in the pro-nuclear world. First I had been surprised by what the advocates were telling me. Nuclear waste, they said, wasn't the worst kind of waste but the best: tiny in volume, contained and monitored, and more aptly called "used fuel" that could be reprocessed and deployed again. Low-dose radiation exposure was not necessarily harmful, they argued; it could even be beneficial. Nuclear power, according to them, had saved vastly more lives than it had taken. Hearing all of these assertions, I felt like I was in Bizarro World.

Then I was surprised when I did my homework and found that these claims, while not uncontested, could find support from credible sources. For example, James Hansen had published a 2013 study claiming that nuclear energy had prevented an estimated 1.8 million deaths that would have otherwise resulted from fossil-fuel pollution.[19] He and his coauthor, Pushker Kharecha, calculated that this number was 370 times greater than the number of lives lost to radiation poisoning and accidents. (Similarly, Kharecha wrote a 2019 paper claiming that if Japan and Germany had kept their nuclear plants open between 2011 and 2017, they could have prevented 28,000 pollution-related deaths.[20]) Time and again, when I was about to write the nuclear community off, I realized that maybe they had a point.

Yet Mothers for Nuclear's support for the Ohio law epitomized what made me uneasy about their approach. Their commitment could sometimes seem single-minded, elevating nuclear energy above more holistic goals like building coalitions with other environmentalists or strategizing a long-term plan for a clean-energy future. I felt this uneasiness when I saw their Facebook posts highlighting the shortcomings of renewables, and when they sounded so certain in their interpretation of what seemed to be an exceedingly complex situation in Fukushima. I understood that they felt they had to fight misconceptions about both nuclear and renewables, but the upshot often seemed to be squabbling with fellow environmentalists, rather than joining them in solidarity against fossil fuel interests.

Sometime in the mid- to late 2010s, a term started circulating on social media to refer to rabidly pro-nuclear people: "nuclear bros." The quintessential nuclear bro was probably Shellenberger, who had celebrated the passage of the Ohio bill full-throatedly on Twitter[21] and had become increasingly hostile to renewables. (Sample headline of one of his *Forbes* columns: "If Renewables Are So Great for the Environment, Why Do They Keep Destroying It?"[22])

But the term was both gendered and implied a degree of abrasiveness that did not apply to all strongly pro-nuclear advocates. This is why I began to think of the more extreme version of pro-nuclear sentiment as "nuclearism," with "nuclearist" as a gender-neutral alternative to "nuclear bro." Eric Meyer was a nuclearist, and the Mothers were nuclearists. They were not at all culturally bro-ish; they were not Twitter trolls; they generally stayed positive and personable. They simply put nuclear energy at the center of their vision of a better world.

Other pro-nuclear organizations, such as the Breakthrough Institute and Third Way, opposed the Ohio legislation.[23] And the decision to endorse it looked even worse a couple of years later, when it emerged that the bill had allegedly resulted from criminal behavior. According to prosecutors, FirstEnergy Corporation—the company that had given Eric funding for canvassing—had spent $60 million to bribe public officials in exchange for passing the bill.[24] It was a huge scandal that reinforced the worst suspicions about the nuclear industry.

Like any "ism," "nuclearism" is an inexact term. But signs that you might be a nuclearist include: supporting virtually any legislation that saves nuclear plants; wearing a hoodie that says "Only U can save the planet" (as Eric Meyer did); having a photo of a nuclear plant as your Zoom background (ditto); and deliberately wearing radioactive jewelry. If you refer to people who oppose nuclear power as "antis," think Amory Lovins is a villain, and use scare quotes around the "spent" in "spent fuel," you are likely a nuclearist.

Through my adventures in the nuclear world, I had come to know the nuclearist outlook pretty well. I was not an expert on nuclear power, but I had become, I daresay, something of an expert on nuclearism. Whenever I spoke with, well, normal people about nuclear energy, and heard the same refrains—*What about the waste? What about Fukushima?*—I felt that I knew exactly what a nuclearist would say in response. I had not become a nuclearist myself. But I was certainly nuclear-curious.

Around this time, when visiting my parents, I came across an oversized yellowing paperback on one of their bookshelves titled *Watermelons Not War! A Support Book for Parenting in the Nuclear Age*. It had a very DIY vibe: it was coauthored by a group of women calling themselves the Nuclear Education Project, and the cover featured a hand-drawn illustration of a watermelon growing on a vine.

As I leafed through it, I learned that it had been written by five mothers in the Boston area in the wake of the 1979 accident at Three Mile Island. "We came together to talk about our hopes and fears for our children in an age of growing nuclear terror and uncertainty," they wrote. "We wanted to find ways to answer our children's soul-shaking questions about the world." Their fears focused on both nuclear power and nuclear weapons, which they saw as two heads of the same monster. "Why are we destroying our life-support system?" they asked in anguish. "Why do we saturate the land, air, and water with poisonous substances that threaten our existence?"

Skimming the pages, I had a complex reaction. As a mother, I related to their feelings absolutely. But the objects of my fears were completely different. I didn't lie awake worrying about nuclear bombs or meltdowns; I was haunted by guns and climate change. And I felt a lick of something resembling guilt. The source of their terror had begun to look like a possible remedy to the source of mine. I felt almost like a traitor to these women with whom I otherwise felt an affinity. I was certain that if I had been a parent in the early '80s, I would have agreed with them wholeheartedly. But now I wasn't sure.

CHAPTER TWELVE

THE TRIBE

THE 12,000 ACRES OF COASTLINE KNOWN AS DIABLO CANYON LANDS, also called the Pecho Coast, is divided up into four parcels. The plant itself and its ancillary buildings and other infrastructure occupy a relatively small footprint—about 585 acres—known as Parcel P. But, presumably to minimize the effects of any potential nuclear accident, PG&E had purchased the surrounding area as well. Down the coast lies South Ranch, where the oak woodlands are located, and then Wild Cherry Canyon. The access road to the plant cuts through this territory. In the other direction, up the coast, lies North Ranch. For the most part, this land has remained off-limits and untouched since PG&E purchased it, with a couple of exceptions: there are two hiking trails with limited public access; and there are some grazing lands where cows nibble at the grasses.

In February 2018, PG&E established the Diablo Canyon Decommissioning Engagement Panel to foster community involvement in decisions about the decommissioning and the future of the site. The panel received applications from 105 members of the public who were interested in serving, and a committee selected 11 of the applicants. One of them was Scott Lathrop, the man who had visited Diablo Canyon while it was under construction, back in the early '80s, the one whose ancestors had lived on the land.[1]

In the decades since that visit, Scott had raised a family in San Luis Obispo and worked as the assistant superintendent of business for two local K-12 education systems. He also did some land development in the area, building 26 homes and a couple of commercial projects. By 2018, his four kids were grown and he was nearing retirement. With his new abundance of free time, he was getting more involved in his tribe, working to advance their interests. At the decommissioning panel's inaugural meeting, in May of 2018 at the County Government Center in San Luis Obispo, Scott sat at the dais with the other panel members. A soft-spoken man with a gray goatee, he leaned into the mic to introduce himself. "I've been in San Luis Obispo for going on sixty-four years, but my family has a history, I want to say, of many moons, and when I say many moons, I'm Native American. I actually have ancestry that came from this particular area. So definitely quite interested in the land use, future land use of the overall site."

Scott had grown up around his extended family who lived in the area—cousins, uncles, aunts, grandparents. They had a tribal council, but they were not a formally recognized tribe. In 2010, though, they were contacted by Chevron, which was planning a project in their ancestral territory and sought their input. As of 2005, California law mandated various types of consultation with local tribes, but to participate, Scott's group would need to form some kind of legal entity. Based on his reading of history, Scott told me in one of our conversations, it was common for tribes to formalize only in response to some kind of external incentive. "And that seems to make sense to me, because most tribes were like tribal families or kinships," he said. "There wasn't a lot of thinking that we had to write everything down and have this rule, that rule." For his group, "There came a point of time when, okay, if we're going to be, quote, heading down that recognition pathway, well, the westerly world was, you know, 'Who are you? What's your identity? How do you operate?' So you kind of have to play the game, if you will, in order to create that standing in the eyes of people today."

A long-term goal was to become federally recognized, but that process can take years or decades. To ensure that they could benefit from the new laws, Scott and his relatives created a nonprofit, which could handle financial matters. The tribal council was chaired by Scott's cousin, Mona Tucker; Scott eventually became the CEO of the nonprofit.

The name "Chumash" had been applied to many tribes throughout California, who spoke languages in the same language family.[2] Scott's tribe had been called the Northern Chumash. But they called themselves the *yak titʸu titʸu yak tiłhini*, which means "the people of the full moon."[3] For the public, when Scott and Mona and their relatives set up their formal entities and their website, they called themselves the *yak titʸu titʸu yak tiłhini* Northern Chumash Tribe—YTT for short.

The state laws requiring tribal consultation represented initial small steps to right historical wrongs. When White men flooded California in the Gold Rush of 1849, thousands of Native Americans were killed with impunity. An 1850 state law, euphemistically titled the "Act for the Government and Protection of Indians," facilitated removing Native Americans from their home territory and separating parents from children. During that decade, the state government encouraged and funded militias to fight Native peoples. In 1851, for example, a state-sponsored militia arrived in Yosemite Valley, shot at the Miwok people living there, and burned their wigwams.[4] The same year, Peter Burnett, California's first governor, predicted that the state would be at war "until the Indian race becomes extinct."[5]

By the late 2010s, pressure was intensifying to do more to make amends. In June 2019, Newsom, by this time serving as governor, issued an executive order formally apologizing to California's Native Americans for this brutal treatment. To honor the occasion, he met with tribal leaders at a wooded outdoor spot in West Sacramento, where a California Indian Heritage Center was planned. "It's called genocide," he said. He also announced the formation of a Truth and Healing Council aimed at correcting the historical record.[6]

The same year, Scott and Mona were pleasantly shocked to learn that the California Public Utilities Commission would be instituting a new rule, the

Tribal Land Transfer Policy. It stipulated that when investor-owned utilities dispose of land, they should try to negotiate a transfer to tribes with historical claims on the property before putting it on the market.

This policy was in the spirit of a campaign brewing among tribal groups known simply as "landback." The campaign was a demand to go beyond well-meaning but fundamentally symbolic gestures such as statue removals and "land acknowledgments" and apologies, and to return actual control over land that was illegally (or at least unethically) taken from Indigenous people. A group called the NDN Collective was demanding the closure of Mount Rushmore and return of the land in the Black Hills of South Dakota. According to the NDN Collective, "LANDBACK is more than just a campaign. It is a political framework that allows us to deepen our relationships across the field of organizing movements working toward true collective liberation."[7] In one of the first victories for the movement, the Esselen Tribe in California would regain control of 12,000 acres of Big Sur in 2020. Four years later, a 125-acre grove of redwoods in Northern California would be returned to the Yurok Tribe.

When it came to Diablo Canyon, the new Tribal Land Transfer policy "really put us, I'd say, in the catbird seat," Scott told me. "In fact, I tell people our stock went up, because everybody that was eyeballing the land for repurposing, they're going, 'Well, shoot, we're gonna have to deal with the local tribe.' All of a sudden, everybody wanted to be our partner."

In 2019, the tribal members began meeting with a few local entities: Cal Poly, the Land Conservancy of San Luis Obispo, and later, REACH, a coalition promoting economic development. Together, they hashed out a vision: ownership of most of the property would be transferred to the tribe, and they would steward it in partnership with the Land Conservancy. After Diablo Canyon was decommissioned, they aimed to establish a "Clean Tech Innovation Park" at the site, which might include desalination, offshore wind, battery storage, aquaculture, and other emerging technologies. It was an inspiring proposal, combining conservation with innovation, forging a cutting-edge future while honoring—and redressing injustices of—the past. The tribe was also able to put together significant funding from private sources, and in 2021, they came to PG&E with an offer. PG&E promised to get back to them with feedback.

The YTT Tribe had a complicated relationship with the nuclear facility. "If we had any say at all about the plant being built on day one, we'd say, you know what? That's probably not a really good idea, because it's not respectful of the land—or the water, because it's on the coast," Scott told me. "But nobody asked us. We had no control." As it turned out, several members had helped build it or later worked there. It had been, on balance, a positive presence in their lives. Scott said that although they didn't have an official position, "I guess if we were ever required or demanded, my guess would be that it would be in favor." But, he knew, this put them at odds with many of their fellow Native Americans.

In the early 1940s, during the depths of war and the intense, secretive activity of the Manhattan Project, White men from a mining company came out to the Colorado Plateau, an expanse of land that includes parts of Colorado, New Mexico, Arizona, and Utah. This area was largely populated by the Navajo people.[8]

Until recently, the Navajo had banned mining on their land. With the outbreak of the war, though, they had pledged their support to the U.S., and did what they could to support the effort. That included allowing mining for vanadium, a metal that helped harden the armor of warships. Mixed in with the vanadium was uranium, but that was considered virtually worthless—until Einstein's letter to President Roosevelt completely changed assessments of its value. Officials leased land from the Navajo and hired, for a pittance, many of the men to work as miners. The miners and their families had no idea what the uranium would be used for, although they were told it was to support the war effort.

By the late 1930s, scientists knew, based on evidence from Europe, that uranium mining was associated with lung cancer, although they weren't sure exactly why. But in the frenzy to build the bomb, little attention was paid to the question, and safeguards were nonexistent. After the war was over, in the 1950s, the U.S. Public Health Service began researching the health risks. They found that in underground mines, miners were inhaling dust full of radon, a radioactive element that was also mixed in with the uranium. Of particular concern were isotopes known as "radon daughters" that could lodge in the lungs. The Public Health Service recommended simple steps that could reduce the health risks, such as ventilation; wetting rocks before drilling to minimize dust; giving the miners respirators; and requiring frequent changes of clothes. But these recommendations were ignored.

There were no standards for cleaning up the mine tailings, which were also radioactive. When they dispersed, they contaminated local groundwater. What was more, some of the Navajo people, uninformed of the dangers, were using the material left over from mining to build their own houses.

In the early 1960s, miners began to suffer from a mysterious disease that turned out to be lung cancer—previously all but unknown among the Navajo. In the following three decades, an estimated 500 to 600 Navajo miners would die of it, while at least as many are estimated to have died since 1990.[9] Other community members also got sick from living in radioactive homes and from the leaked tailings. Their organizing eventually resulted in a 1990 federal law, the Radiation Exposure Compensation Act, which provided for compensation to miners, as well as to "down-winders" who were exposed to fallout from weapons testing.

This episode is perhaps the most famous and egregious example of how Native people have suffered as a result of uranium exploitation. There are others. On July 16, 1979, just four months after the partial meltdown at Three Mile Island had riveted the world, a massive spill of uranium tailings occurred at the Church Rock uranium mill in New Mexico. The mill was bordered by Navajo lands. The accident released far more radioactive material than the accident in Pennsylvania, and in fact, more than any other accident in U.S. history. But it received very little notice at the time, and remains obscure.

Most of the early mining in the U.S. was for weapons, not the civilian reactor program. (As the civilian reactor program grew, imports increased from countries such as Australia and Canada, where higher-quality uranium ore had been discovered.) Still, some of the uranium mined in the Colorado

Plateau had been used for civilian reactors, and uranium mining was a necessary step for nuclear energy production. While public attention focused on the risks of reactor accidents and radioactive waste, more health harms were occurring earlier in the fuel cycle, disproportionately affecting Indigenous people. All of this helps explain why Native Americans have, on the whole, been deeply suspicious of nuclear power.

In 1996, a coalition of Indigenous groups convened the Indigenous Anti-Nuclear Summit and issued the following declaration: "We, the Indigenous Peoples gathered here for this summit, standing in defense and protection of our Mother Earth and all our relations, do hereby unanimously express our total opposition to the nuclear power and weapons chain and its devastating impacts and deadly effects on our communities."[10]

Not all Native people were unequivocally opposed to nuclear energy. Over the years, some tribal leaders were open to mining and milling, provided that it yielded economic benefits and they could have some control over how it was carried out. (Today, very little mining or milling occurs in the United States, although some efforts are underway to revive both.) The whole controversy raised questions familiar from the nuclear debate: are the problems we have seen inherent to the nature of the technology, of the material, or are they rooted in historical circumstances and susceptible to improvement?

Scott believed the latter. He saw the impacts from mining and milling as yet another example of how the U.S. government had betrayed Native Americans, but not as an indictment of nuclear energy itself. Over time, in fact, he came to be a strong supporter of nuclear.

Still, what he ultimately cared most about was the land. On the Decommissioning Engagement Panel, he and his fellow members discussed what the future would hold for the Pecho Coast. Some members felt very strongly about public access for everyone to enjoy the land. With Scott's input, the panel issued a "vision document." It included this compromise statement: "The request for land ownership by the local Native American community should be acknowledged and considered as a valid claim for historical reasons, while bearing in mind the overwhelming public testimony that the Diablo Canyon Lands be conserved and available to the public for managed use."[11] There would still be challenges and stumbling blocks, but the YTT Northern Chumash seemed closer to getting their land back than they could have imagined just a few years earlier.

CHAPTER THIRTEEN

"BUT WHAT ABOUT THE WASTE?"

On a cool, cloudy Friday morning in June of 2023, I once again found myself at a nuclear plant, this time at SONGS, the shuttered facility in San Clemente I had seen from some distance when walking on the beach with Eric Meyer. Driving from the north, I glimpsed the two containment domes—known locally as "the boobs"—ahead of me, just off the freeway. After parking in an adjacent lot, I walked over to a table where I was given a hot-pink vest, safety glasses, and a badge to wear around my neck.

Security was more lax than at Diablo Canyon, given that the plant was no longer operational. SONGS was in the lengthy process of being decommissioned, an almost unfathomably elaborate undertaking that had begun in 2013 and was expected to continue until at least 2028. It involves dismantling the entire facility and shipping away all of the constituent parts on rail lines constructed for the purpose. By the time of my visit, more than two dozen buildings had been demolished and mountains of debris had been loaded onto rail cars and shipped away. Millions of pounds of metal and steel and tens of thousands of titanium tubes would eventually follow, some to be recycled, some to landfills. Some of it was contaminated and considered low-level radioactive waste. This would be shipped to an underground repository in New Mexico. Eventually, the two iconic reactor domes would be dismantled and removed as well. Ultimately, if all went according to plan, the land would revert to its previous state, with no perceptible trace of the nuclear plant left, the residual radioactivity reduced to a level deemed safe by the NRC. Then the property would be returned to the Navy, which owned the land.

But there was one serious obstacle in the way of complete decommissioning. Even as all of that debris and dismantled equipment was transported to other destinations, there were no plans to relocate the most toxic material of all: the high-level radioactive waste that had accumulated over the course of the plant's operations, from 1968, when the plant opened, to 2013. A total of 1,600 tons were stored in 123 steel canisters on a concrete pad overlooking the ocean. There were no plans to move it because there was nowhere for it to go.

The occasion for my visit was a press conference by Mike Levin, the congressman who represented the district, and Jennifer Granholm, President Biden's secretary of energy. Levin seemed like someone who was always destined to run for office—in a good way. The son of a Jewish father and a Mexican American mother, he grew up not far from the plant. He had gone on to graduate from Stanford and Duke Law School, then returned to Orange County to practice environmental law. He was both wonky and folksy, approachably handsome, and basically impossible not to like. I wasn't surprised when I later learned that he had been urged to run numerous times before finally deciding to do so after Trump was elected in 2016.

Since taking office in 2019, Levin had made moving the spent fuel from the coast one of his top priorities. He had put together a task force to study the situation and issue recommendations, inviting pretty much anyone and their brother in the community to join it.[1] He'd cofounded the bipartisan Spent Nuclear Fuel Solutions Caucus, and he'd helped secure $93 million in Congressional funding to work on these solutions. He had also sponsored a bill called the Spent Fuel Prioritization Act, which would give precedence to relocating the spent fuel from the riskiest sites.

SONGS was at the very top of that list: it was in a densely populated area (9 million people living within 50 miles, as Levin frequently pointed out); near multiple earthquake faults; and only 100 feet from the sea, in the context of rising sea levels and coastal erosion. Although Levin was a progressive Democrat, the issue seemed, to an unusual degree in 21st-century America, to transcend partisanship; his cosponsor on the bill was Darrell Issa, a conservative Republican who had formerly occupied Levin's seat and then subsequently won a race in a neighboring district. (Issa had also bankrolled the recall campaign against Gray Davis.) "Nuclear waste doesn't care if you're a Democrat or a Republican," Levin liked to say.

This cloudy day in June, Levin and Granholm would get a photo op in front of the pad where the waste canisters were stored and make what was teased as a big announcement. Wearing my hot-pink vest and safety glasses, I entered a turnstile gate and walked down a sloped paved road to a landing where a podium and microphone were set up. I milled around, chatting with the other members of the media who had shown up. There were a couple of local TV stations, a photographer from *Agence France-Presse*, a reporter for the local *Coast News*. For some reason, this event at a radioactive waste dump had attracted only a handful of journalists.

We were, in fact, some distance from the concrete pad—technically known as an independent spent fuel storage installation (with a melodious acronym: ISFSI, pronounced "Isfuhsee"). But we could see it, between us and the ocean. Down below us were several rail lines with a couple of immobile rail cars on them. Behind us and to the right were bluffs that appeared to be covered in brownish material, although I couldn't quite tell if it was a natural part of the landscape or not. When I asked the media relations person, she told me it was gunite, a mixture of sand, water, and cement. It had been applied to stabilize the bluffs, and would eventually be removed. In some of the gunite, a profusion of grasses and wildflowers had sprung up. It was one of my favorite phenomena, almost the opposite of spotting a plastic bag or a beer can in the wilderness: seeing how nature blooms in the most industrial, artificial sites. Seagulls squawked and flew by. Out in the water, a couple of surfers in black wetsuits were riding the waves.

At last, Levin and Granholm emerged and took their positions behind the podium, which bore a red, white, and blue sign reading, "President Joe Biden" and "Investing in America." Levin, wearing a navy suit jacket and striped tie, spoke first. "It's a big day, because for the first time in over a decade, we're here to announce that the federal government has a plan when it comes to managing the nation's spent nuclear fuel," he said. "I've been in Congress for five years, and for the first time, we finally, finally have a plan."

This would turn out to be something of an overstatement—it was more like a plan to make a plan. But still, even that was progress.

OF ALL OF THE CLASHING PERSPECTIVES HELD BY NUCLEARISTS AND ANTIS, perhaps no difference is as stark as the one over nuclear waste. For many opponents, it's the single aspect of nuclear power that they simply can't countenance. For supporters, the waste—or used fuel, as they prefer to call it—is one of the energy source's most winning features.

The case against nuclear waste is pretty simple. As the uranium atoms in the reactor split apart, the process creates new radioactive elements, including plutonium. Being in the same room with unshielded nuclear waste, fresh out of the reactor, would give you a lethal dose within minutes. (It might take you a while to die, but that's not necessarily a plus; it would be a long, agonizing death.) According to a landmark 1957 report from the National Academy of Sciences, "Unlike the disposal of any other type of waste, the hazard related to radioactive waste is so great that no element of doubt should be allowed to exist regarding safety."[2]

The waste also intersects with other areas of concern about nuclear power. The fact that it contains plutonium means that it's possible to use supposedly peaceful civilian reactors to create bombs. In 1974, India, for example, tested an explosive device that had been built with reprocessed spent fuel from a research reactor.[3] (This has rarely occurred, however, and there are easier ways to develop a weapons program.) The spent fuel can also be involved in accidents, especially when it's still in cooling pools, not yet transferred into dry storage. As we've seen, a major concern in Fukushima was that the spent fuel pools would lose their water, which could have led to a fire and a large radioactive release.

The case for nuclear waste is largely a matter of putting it in context. Stewart Brand covered most of this in his book: the waste volume is tiny compared to that generated from other sources; it loses toxicity over time. Of course, this process takes a long time, but the nature of radioactivity means that the "hottest" elements decay at the fastest rate. The longest-lasting elements are the least radioactive. According to one estimate, after about 500 years, the radioactivity will have declined by a factor of 50,000, and it wouldn't be dangerous unless you actually ingested it.[4] It could still pose health risks if it came into contact with rivers or groundwater, but no more than myriad other toxic substances.

And it has the potential to be recycled. When it's removed from the reactor, more than 90 percent of the energy remains untapped. The reason it's removed is that the fission process generates "nuclear poisons" that eventually get in the way of further fissioning. So, every 18 months or so, reactors undergo planned outages while workers replace the fuel. It is not easy, but it is possible to reprocess this used fuel and use it again. The U.S. experimented with reprocessing for a time, but in 1977, President Carter placed a moratorium on it. His concern was that the material could be diverted for weapons production—as had recently occurred in India. The goal was apparently to set an example to other countries, but several, including France and the UK, went ahead with reprocessing anyway. In 1981, President Reagan lifted the moratorium, but interest in, and capacity for, reprocessing remained low (though that has recently begun to change, as we will see).

As nuclearists see it, our current system for storing waste, far from being a crisis, is fine—great, even. The used fuel is isolated, contained, and monitored (which has become a kind of mantra), and there are no reported cases of harm that it's caused.[5] When it's stored—first in cooling pools, then in dry casks—the radiation is blocked with water or steel and concrete, so that virtually none escapes. How dangerous it is, then, perhaps depends on how you look at it: whether you consider its sheer potential to cause harm or its record of actually causing harm—i.e., what could happen versus what has happened. This is in contrast to the fossil fuel pollution from vehicles and power plants that gushes freely into the air and the plastic that chokes rivers and suffuses the ocean. As with nuclear power generally, you could argue, spent fuel is safe because it's dangerous: knowing that it posed a more acute risk, we have developed systems to prevent it from reaching people and harming their health.

As a result, nuclearists see used fuel as an asset more than a burden. As one advocate, Madison Hilly, wrote in a 2023 *New York Times* guest essay, "I feel very comfortable with the way we manage nuclear waste, with making more of it and with passing this responsibility on to our kids. I hope my daughter's generation will inherit many new nuclear plants making clean power—and the waste that comes with them."[6] A couple of months later, she posted on Twitter a photo of herself pressing her pregnant belly up against a waste storage cask.[7]

In the United States, there are about 80 locations in 35 states where high-level radioactive waste is being stored indefinitely—mostly at operational and retired nuclear plants. In terms of what to do about it, the expert consensus, dating back to that 1957 National Academy of Sciences report, has held that we should ultimately bury it deep underground in a repository. Anti-nuclear people, I found, generally agree with that goal, but are pessimistic about its prospects for becoming a reality. The most strongly pro-nuclear people, in contrast, tend to think along the lines of Stewart Brand: a repository is pointless; the spent fuel is fine where it is; and it would be better to keep it accessible for possible reprocessing than to inter it. Some nuclear proponents, though, do support the long-term goal of a repository. Even with reprocessing, there would eventually be a residual waste product. And even if they themselves aren't worried about the current system, they know it would be good to have a more definitive answer to the inevitable question, "But what about the waste?"

Yet in the U.S., plans for a repository completely broke down almost a decade before Congressman Levin took office. Since 1987, the intent had been to bury all of the nation's high-level radioactive waste at Yucca Mountain. But resistance in the state was strong—it became known as the "Screw Nevada" law. In 2010, the Obama administration announced that it would withdraw the license application to the NRC for the repository. "We're done with Yucca," said President Obama's energy advisor Carol Browner.[8] By this time, about $9 billion had been sunk into the project.[9] It was considered one of the most extensively studied sites in the world, and more than five miles of tunnels had been drilled in the ridge. But Nevada's powerful long-serving senator, Harry Reid, made it his mission to see that the repository would never come to pass. It was widely believed that Obama appointed Greg Jazcko, a Reid ally, chairman of the NRC in order to help kill the project.

Meanwhile, Obama's secretary of energy, Steven Chu, convened the bipartisan Blue Ribbon Commission on America's Nuclear Future to devise a new strategy. In 2012, the commission issued a report, which stated, "The overall record of the U.S. nuclear waste program has been one of broken promises and unmet commitments."[10] The report's recommendations included developing one or more permanent underground repositories as well as one or more consolidated storage facilities (that is, centralized interim sites where spent fuel would be stored aboveground). But its first recommendation, highlighting its importance, was a "consent-based approach" to siting both. The fatal flaw of the Yucca Mountain plan had been that it was imposed on Nevada; Nevadans had no say.

This approach was based on observations of more successful programs in other countries, including Finland and Sweden. The basic idea was that communities should be respectfully engaged from the outset. Only communities that volunteer should be considered. Then they should be invited to participate in dialogue and negotiation. People should ideally be compensated for time spent in this process—showing up at meetings and hearings—and they should receive funding to hire independent experts for information and advice. The negotiations should include some sort of benefits package based on the community's specific needs and preferences, whether that means job guarantees, direct financial compensation, or amenities such as playgrounds. And they must have the right to say no at any point in the process.

The previously standard approach to siting nuclear projects—whether waste facilities or plants—had sometimes been called "Decide, announce, defend." The new approach would be, in the somewhat jargony, touchy-feely language of the Blue Ribbon report, "adaptive, staged, and consent-based."

The term "consent" evokes questions that are uncomfortably reminiscent of the word's more familiar usage. By definition, it seemed inevitable that only communities with fewer resources and advantages would sign up, because wealthy communities would not be in need of such benefits. And just as power imbalances between individuals complicate notions of consent in intimate relationships, so do power (and wealth) deficits in this context. Some believe that it is impossible for a community to meaningfully consent to hosting radioactive waste.

That question aside, this process would obviously take time; that was part of the point. It would be challenging to get a community to agree—and to even determine how "consent" would be defined, since there would inevitably be a range of opinions. But even if a community did step up, other stakeholders, most importantly state officials, would also need to be on board. And although support for dealing with the waste was ostensibly bipartisan—nobody objected in principle—opposition in practice, at the state level, tended to be equally bipartisan. There were so many bottlenecks and veto points in the process of moving the fuel that it was seemingly impossible to make progress. But it was clear that the alternative—forcing the waste on a community—was not only unethical but destined to fail completely.

In nuclear circles, the Blue Ribbon Commission report was considered seminal, laying out a road map for action. In the following years, however, approximately nothing happened. There were too many competing crises, from police violence to climate change to the rise of Trump. As a rule, Americans were not calling their representatives about nuclear waste, or casting their votes based on the issue. There were some exceptions to this rule, though. And nearly all of them, it seemed, lived in the vicinity of San Onofre Nuclear Generating Station.

In February 2022, I met Marni Magda on a crisp, sunny day outside her house in Laguna Beach, about 20 miles north of the nuclear plant. Laguna Beach is an artsy town where the main drag is lined with galleries, cafés, and upscale boutiques, and it's common to see pedestrians walking barefoot in their swimsuits on the sidewalk. A retired elementary school teacher in her mid-70s, Magda seemed in many ways to lead a charmed life. For decades, she's lived in a home just blocks from some of the most spectacular coastline in California; it's a modest cottage, but today would certainly be out of reach for a middle-class family. Her two sons live nearby, and she spends much of her time with her grandchildren, surfing and playing on the beach. But the waste languishing 20 miles away haunts her; it's like a blight on paradise.

Although many of her generational peers were anti-nuclear activists back in the '70s and '80s, she never gave much thought to nuclear power until March 2011, when the meltdowns in Fukushima occurred. One day not long afterward, she was reading *Scientific American* magazine in her living room when she came across an infographic titled "Aging Fleet under Review." The two-page spread showed a map of nuclear facilities across the United States, with color coding to indicate seismic risk. Looking to Southern California, she spotted a nuclear plant icon superimposed over an alarming shade of red. It represented SONGS, which was still operational at the time. "I realized as a California resident, we were in a great deal of trouble," she told me. For Magda, the interpretive map acquired a status akin to a sacred text; she kept it and had it laminated. She handed it to me as we sat at a table in her backyard, birds chirping riotously in a large orange tree heavy with fruit.

Soon after this wake-up call, she'd joined her local chapter of the Sierra Club, and found other community groups who also feared another Fukushima. There were quite a few of them: San Clemente Green, the Samuel Lawrence Foundation, Public Watchdogs, San Onofre Safety, Residents Organized for a Safe Environment. She began attending protests in the state park adjacent to the plant, marching and holding banners aloft.

In 2013, when Southern California Edison announced that the plant would be decommissioned because of the faulty steam generators, many community members hailed the decision as a victory. "Oh, I was thrilled," Magda told me. "I was absolutely thrilled." The celebration, however, was short-lived. Realizing that all of the accumulated high-level radioactive spent fuel would remain at the site for the foreseeable future, they shifted their focus to the waste.

Since the plant shut down, Magda has spent pretty much all of the time she's not frolicking on the beach with her grandchildren on the problem. She wants to get the spent fuel moved to a more suitable site and to ensure that, until then, it is stored as safely as possible. As of 2017, she has been the Sierra Club's representative on the Community Engagement Panel, an entity established by Edison that holds quarterly meetings with the public. In her garage, she showed me a gray filing cabinet with four vertically stacked drawers, her granddaughter's teal bike propped up against the side. The drawers contain hundreds of manila files with hand-scrawled labels: "Western Interstate Energy Board," "Sierra Club—Comments on Scoping Process," "Thick vs. Thin Canisters & Corrosion Cracking." Inside the folders are notes from meetings, business cards of public officials, news clippings, academic papers, and government reports.

Magda has traveled with other members of the Community Engagement Panel around the country to learn about different canisters that might be used to store the fuel on-site. Most surprising to me, she also has spent plenty of time on the site itself. When I asked if she had ever been on a tour, she said, "I used to lead them." As worried as she is, she thinks the waste is stored as securely as possible for now, and that on a day-to-day basis, it is safe. Her anxieties revolve around events like earthquakes or terrorist attacks, or, over the longer term, coastal erosion and sea level rise. "We can't leave it there," she told me.

Edison had established the Community Engagement Panel to try to build trust with residents, many of whom were deeply suspicious of the utility. The meetings were well attended, and before they shifted online because of the pandemic, often raucous. David Victor, a professor at UC San Diego whom Edison had tapped to lead the panel, told me that at some meetings, people had to be escorted to their cars. "Folks were just out of control." When the meetings switched to virtual because of the pandemic, they became much more civil. A mute button will do that.

Usually, Victor told me, the main response when a nuclear plant shuts down in a community is disappointment over the loss of jobs. But this area was different, perhaps in part because it was more populated. The population of San Clemente had more than tripled since the plant began operations in 1968, from about 17,000 to almost 65,000; the population of the broader Southern California region had of course also soared. The plant was a major employer, but it was not the primary engine of the local economy, as it was in some remote communities. A big part of the economy was coastal tourism, and if anything, the plant was seen as a liability on that front. "The ocean is everything in this community," Levin told me when I interviewed him at his local office. "We are so much a product of our ocean, whether it's recreation, whether it's the surf culture, which is a very real thing." Plus, it was anti-nuclear California.

I spoke with a number of the local activists, some from the Community Engagement Panel and some who thought it was a farce ("Community Enragement Panel," the latter called it). I soon realized that even among those who were most anxious about the waste, there were clashes over the best way forward. One faction, including Magda, wanted to find an interim storage site somewhere else in the country and move it there as soon as possible. But another thought it should be moved just across the freeway to the Navy's land there, where at least it would be on higher ground, farther from the ocean. Sarah Brady, a young surfer (and third-generation member of San Onofre Surfing Club), told me, "We'd be dumping it on low-income communities that don't want it. . . . And it's our waste and our responsibility. We are an affluent coastal community that's benefited from this power." She added, "Also, these short-term sites would likely become de facto long-term sites." The problem with the proposal to move it across the freeway, though, was that the Navy had flatly said no.

I also spoke with Rob Howard, who represented Oceanside (another nearby beach town) on the Community Engagement Panel, and had worked at SONGS for 31 years. He had a distinctive perspective on energy issues and the nuclear industry. Toward the end of his time at the plant, he'd founded a consulting firm focused on microgrids with renewable energy, where he now worked full-time.

We met at a café in Irvine. He had just come from a meeting with the Orange County Power Authority, a new entity whose purpose was to offer residents a community-based, renewables-driven alternative to Edison, so he was wearing a sober gray suit. Over a turkey sandwich and apple juice, he told me about his time at the plant. "I did pretty much everything," he said. That included working as a reactor operator and staying on to work on the decommissioning. He was an extrovert with a commitment to public service he'd inherited from his parents; his father was a Baptist preacher, "but of course my mom ran the church," he said. In addition to his job at the plant, he was also heavily involved in the union. Among other roles, he served as the labor representative to the California Public Utilities Commission's diversity council.

Howard was one of the few Black employees at the plant. At work, as in the rest of his life, he explained, he was obliged to put himself "in a position where people aren't afraid of me. I cross the street before you do because I don't want you to accuse me of anything. I can't get angry at work. The onus is on me to keep you from being fearful."

He described one incident he had at the plant years ago that had stuck with him. A supervisor had instructed him to wait in a room. "So I'm sitting there, waiting for the supervisor to come back." Then one of the reactor operators came back and summoned him to help with something. "I said, 'Hey, he told me to stay right here.'" The response: "Profanity-laden tirade. Cursed me out. I'm lazy, I'm this, I'm da-da-da. I'm standing there kinda in shock. . . . Did he just . . . ?"

As he was reeling, two other operators grabbed him and moved him out of the room. "And I said, 'What the hell was that?' Their comment was, 'We was worried about what you might do to him and we didn't want you to get in trouble.'" The actions of his purportedly well-intentioned colleagues were more upsetting to him than the initial abuse. "Now this dude just cussed me out, went off on me. And they moved *me.* So while people try to say that race doesn't matter and I don't see race and all that other stuff, that's garbage. If I can tell you this phone has got a black case on it"—he patted his device—"I see color."

As I was absorbing this story, he went somewhere with it that I did not anticipate. He said, "Nuclear power, I kind of had a kinship with it, because I knew people were afraid of it, and I knew what that felt like."

I was struck when he said this, because it reminded me of comments I'd heard from White pro-nuclear advocates, although they wouldn't dare to explicitly liken anti-nuclear sentiment to racism.

Howard wasn't a nuclearist; his current work focused on renewables, after all. But when I asked later whether he thought we still needed nuclear, he said, "I don't see anybody getting a smaller phone, a smaller TV." He noted that part of the process of decarbonization is electrifying more energy uses, meaning that we'll need more electricity in the future for, say, cars and heat pumps. "I would include it as a part of the portfolio because I need baseload."

He went on, "If you actually get to know it, you think, wow, we've got this much power, no greenhouse gas emissions." He supported efforts to relocate the waste at SONGS, but he wasn't overly worried about it. "Yes, we got radioactive waste, but we know how to deal with it." And he reiterated his earlier point. "The reason I embraced nuclear power was because I felt it was like being Black in America. People think they know you but they don't. And they're afraid of you."

CONTRARY TO THE POPULAR IMPRESSION OF NUCLEAR WASTE AS GREEN goo, spent fuel consists of solid pellets, each about the size of a thimble. At SONGS, the fuel rods, bundled into fuel assemblies, were retrieved from the reactors over the course of years. After cooling for at least five years in pools of water, they were transferred into stainless steel canisters. Workers then transferred the loaded canisters, at 50 tons each, out to the concrete pad overlooking the ocean and lowered them into stainless steel cavities beneath the pad.

Months before Levin's press conference, I had taken a tour of the waste storage site, the ISFSI, itself. On a bright, gusty day, I was led by John Dobken, Edison's public information officer, and Jerry Stephenson, the ISFSI engineering manager, through security and outside to the concrete pad. On our way, we passed workers in hard hats engaged in the ongoing demolition of the plant. Out on the concrete pad, in my immediate vicinity, I could see a partially razed industrial site; farther out, it looked like a commercial for a seaside vacation. The pad is elevated above sea level, but the ocean is about 100 feet away, on the other side of a seawall. "Dry fuel storage is very self-sufficient," said Stephenson. "There's no fans, there's no cooling systems, it just sits there, fat, dumb, and happy, as I like to say." Pointing out some decoy birds of prey flapping around in the wind, he told me they were intended to deter seagulls. His biggest headache, he said, was seagull poop on the concrete enclosures.

We could not see the canisters, which were below our feet, inside the enclosures. Dobken told me that each canister was surrounded on the sides by ten feet of concrete, as well as a three-foot reinforced concrete slab above and below, all of which absorbs radiation that isn't shielded by the stainless steel. According to Edison, the canisters were designed to withstand any earthquake that could occur in the area and up to 125 feet of water.[11]

I mentioned that I felt warm air emanating from the enclosures. Stephenson explained that vents allow air to continuously cool the spent fuel, and the air comes out heated. "I joke that we should put a little restaurant out here and serve drinks," he said. But he assured me that almost no radiation was escaping. Local activists had successfully advocated for the installation of a radiation monitor on the pad. We went over to look at it, and it showed a reading of 13 microrem per hour, just slightly higher than the area's natural background radiation. (For context, over the course of a cross-country flight, passengers are exposed to about 3,500 microrem.)[12]

I wanted to come to some sort of definitive conclusion about how scary the waste was. But as always with nuclear power, that kind of answer felt elusive. This location was clearly not ideal, I thought, looking at the waves lapping the shore nearby; it would take fewer than 20 me-sized people lined up head to foot to reach them. And I didn't believe assurances that no tsunami above certain dimensions would ever occur here, or that the casks were guaranteed to remain unaffected by any number of events. I didn't buy them not because I thought Edison was lying, necessarily, but because if recent years have taught me anything, it's that our models and predictions are never perfect or stable. (Before 2011, authorities had not anticipated a tsunami of that size off the coast of Japan, either.)

At the same time, I couldn't seem to get worked up about the risks. I couldn't imagine ranking the waste as my top concern, even though I lived within the 50-mile radius Levin was always invoking. It wasn't even my top concern about the Southern California coast. In October 2021, a broken pipeline had resulted in an oil spill covering more than 8,000 acres of the ocean surface, including in San Clemente. It affected beaches I visit with my family. Meanwhile, the ocean was rapidly warming and acidifying. In terms of public health and safety risks, there were too many others to count. In October 2015, a natural gas storage facility in Los Angeles sprang a colossal leak that was not plugged until the following February; along with an estimated 97,100 metric tons of planet-warming methane, it spewed the carcinogen benzene and the neurotoxin n-hexane into the air. Every summer and fall, we breathe in smoke from increasingly destructive wildfires. Most of these risks have something to do with fossil fuel combustion, although there are many others I could list as well.

Certainly, it seemed possible that something bad could happen at some point as a result of the waste. But, based on the reading and interviews I'd done up to that point and seeing for myself how securely the waste was stored, it seemed like the worst-case scenario would be some moderate release of radiation—nothing explosive or dramatic. The world we inhabit is now so far from pristine that to me, I guess, nuclear waste was no longer uniquely terrifying.

Still, I was not keen to linger longer than necessary on the pad. Before we left, Stephenson asked if I wanted to get a photo of myself standing on top of one of the enclosures.

"I'm good," I said.

At the press conference, after Levin finished speaking, Secretary Granholm took the podium. She was petite, with stylish short blond hair and a coral jacket. "Let me just start by saying that this administration is committed to ensuring that nuclear power remains part of our carbon-free energy mix," she said. "But we recognize that in saying that, we all have to figure out the solution about the waste issue. And that is not a solution that you can impose on any community."

And then she made the big announcement. Thirteen entities had been selected to collectively receive $26 million to serve as "consent-based siting consortia." These consortia consisted of universities, nonprofits, and nuclear industry firms who had applied to participate in discussions with communities about how to manage the waste. I was intrigued to learn that Mothers for Nuclear and Scott Lathrop of the YTT Northern Chumash Tribe were both taking part in one of the consortia.

They were not yet at the point of looking for a permanent underground repository. They were not even yet at the point of selecting sites for temporary storage. These groups were merely tasked with starting conversations, to try to figure out effective ways of engaging with local communities. It was baby steps, by design.

As such, it reflected a larger shift in nuclear culture: an attempt to learn from past mistakes and adapt to the 21st century. You couldn't foist a nuclear project—whether a repository, an interim storage site, or a new reactor—on people. They had to believe that it would benefit them. They had to welcome it. This would be the new nuclear ethos. When, and whether, it would actually result in any new projects remained to be seen.

CHAPTER FOURTEEN

THE GUY IN THE HEADBAND

Every Thursday at 6:00 p.m. in downtown San Luis Obispo—unless it's raining, which it usually isn't—the weekly farmers' market begins. Five blocks of Higuera Street are closed to traffic, and the vendors set up their tents and their wares. The market, which dates back to 1983, is a local institution that has acquired the vibe of a street fair. If you were to make your way through the throng, you would see shoppers poring over plums and squash and kale, palpating tomatoes and avocados. But you would also see the Udderly Awesome ice cream truck and children running around with painted faces or jumping in a bounce house. You might see aerial acrobats climbing silks, a clown juggling, a man in a bear suit giving out hugs. And, at a green tent in front of Mother's Tavern, between Broad Street and Garden Street, you would almost certainly see Gene Nelson.[1]

Gene is a sturdy man with very little hair except for a long, surprisingly luxuriant gray ponytail, and he nearly always wears an elastic green headband around his head. The sign on his tent reads "Californians for Green Nuclear Power," which might be confusing for a moment. Is the point to assert that nuclear power *is* environmentally friendly? Or is the implication that it needs to become green? If you paused at the booth, the answer would soon become clear. Gene would try to give you a handout about the benefits of Diablo Canyon, California's "climate champion," or ask you to sign a petition to show your support for the plant. In his rather mellifluous voice, with the patience of a teacher, which he was for many years, he'd answer any questions you might have about Diablo Canyon or nuclear energy in general, the same questions he's answered countless times. If you brought up seismic concerns, he'd tell you that if an earthquake did strike, the place he'd want to be is Diablo Canyon, because of its structural soundness. If you were to mention waste, he'd retort that the number of people killed or harmed by spent nuclear fuel is a round number: zero. The conversation might be hard to get out of gracefully.

After the plan to close Diablo was finalized in 2018, members of Friends of the Earth and NRDC and the other parties to the proposal exulted. "Diablo will be the first nuclear power plant retirement to be conditioned on full replacement with lower cost zero-carbon resources," wrote Peter Miller, NRDC senior scientist, in a blog post on the group's website.[2] Even Jane Swanson and other members of Mothers for Peace, who had wanted the plant to close earlier, were on balance glad that the agreement had been reached, that the plant's days were numbered, that the long fight over Diablo Canyon was over.

Gene Nelson was among the quixotic group of people, a motley crew of nuclearists, who refused to accept that fate. Like most nuclearists, he had a conversion story. He told me his when we met for lunch at the Madonna Inn, a beloved, famously gaudy San Luis Obispo hotel and restaurant just off the freeway. (Gene encouraged me to check out one of the main attractions, the restrooms: the women's room has chandeliers and fuchsia metallic wallpaper; the men's boasts a urinal with a motion-activated waterfall.) He was wearing not only his green headband but also a green Californians for Green Nuclear Power T-shirt—with a white sketch of Diablo Canyon emblazoned on the back—and green shorts. In the spacious, pink-themed dining area, we sat at a table next to a window. Over a mug of split pea soup with ham, he explained that in the 1970s, he'd marched against nuclear power. At the time, he was a graduate student in radiation biophysics at SUNY Buffalo. He was influenced by a mentor who had traveled to Hiroshima in the medical corps to survey the city in the aftermath of the atomic bombing and, as a result, became an ardent opponent of nuclear weapons. Gene became one, too, and he had the impression that civilian nuclear power plants were essentially a front for making bombs.

In time, several factors changed his mind. In graduate school, he had learned the theories of Hermann Muller—that even the smallest exposure to radiation was harmful. Much later, in the 2000s, he began to read about new research on radiation suggesting that, below a certain threshold, cells can repair any damage they suffer as a result of exposure. One emerging school of thought, led by a toxicologist named Edward Calabrese, even posited that low doses could trigger a beneficial effect, known as hormesis. Radiation science remains highly contentious, with ongoing debates over these more recent theories, but Gene came to believe them wholeheartedly.

He also came to invert his view about the relationship between nuclear energy and nuclear weapons. His initial assumption was not entirely baseless—civilian reactors can be a front for making bombs, as in the case of India. But in most cases, of course, they simply generate electricity. And Gene heard about a program, a phenomenally successful but largely unsung one, that had done essentially the opposite. It had turned nuclear weapons into electricity—the ultimate "swords into plowshares" maneuver.

The program had come about in the wake of the Cold War. On October 24, 1991, Thomas Neff, a physicist who was affiliated with MIT's Center for International Studies, published an op-ed in the *New York Times* headlined "A Grand Uranium Bargain."[3] At the time, the Soviet Union was disintegrating, and Neff was deeply worried about the fate of tens of thousands of nuclear weapons scattered throughout the region. The former empire had been thrown into chaos, and the government was bankrupt. If those guarding the nuclear materials were not paid, Neff feared, they would have both less incentive to do their jobs and more incentive to sell the warheads to the highest bidder, whether terrorists or "rogue states." The stakes could not have been higher.

Neff, an expert on the international uranium market, had an ingenious idea: Russia could dismantle weapons, extract the highly enriched uranium, "downblend" it, and sell the low-enriched uranium to the U.S. to use as fuel in American nuclear plants. Russia would get an infusion of badly needed cash, the proliferation risk would be much reduced, and the U.S. would secure fuel whose uranium had already been mined. This kind of conversion had never been attempted before, but based on his knowledge of the physics, Neff was confident that it would work.[4]

After publishing the op-ed, Neff labored to make this proposal a reality. In August of 1992, President George H. W. Bush announced that an agreement had been reached, and President Clinton officially initiated the program, known as Megatons to Megawatts, in 1993. Over the course of 20 years, the U.S. paid Russia about $17 billion to destroy more than 20,000 nuclear weapons. Amazingly, the resulting fuel supplied 10 percent of U.S. electricity during this period—all electricity, not just power generated at nuclear plants.[5] The program lapsed in 2013, but Gene nursed hopes that something similar could be revived and that, ideally, the U.S. could do the same with its own nuclear warheads. "Nuclear power reactors are the only way to absolutely and positively render nuclear material unusable for bombs," he told me between spoonfuls of soup.

More so than some nuclearists, Gene was quick to suspect conspiracies. He tended to believe that fossil fuel companies were behind most efforts to promote renewables—because they knew that nuclear was their only real competition, he said, and that attempts to shift to 100 percent renewables would leave their position largely intact. He pointed me to advertisements for the natural gas industry that touted their product as the perfect partner to renewables. Natural gas ramps up and down much more readily than other power plants, including nuclear, which means it can easily accommodate the fluctuations associated with wind and solar. So, in that sense, it does complement renewables beautifully; the problem is that it spews methane, a greenhouse gas far more potent than carbon dioxide.[6]

Gene singled out his contemporary Amory Lovins, the renowned renewables champion, for particular contempt. He liked to cite a 2008 *Democracy Now!* segment in which Lovins had said, "I've worked for major oil companies for about thirty-five years, and they understand how expensive it is to drill for oil."[7] (When I asked RMI, the organization Lovins cofounded, about this, a representative gave me this comment: "Amory Lovins has never been an employee of an oil company or any other for-profit company. From 1982-2019, as co-founder and chief scientist at RMI, he consulted for the oil industry–as he has for clients of many other sectors–and his advice has been consistent: to get out of the hydrocarbon business.") Gene called him "nothing but a whore."

In 2006, Gene had moved to the San Luis Obispo area, where he taught as an adjunct professor at Cal Poly and then at a nearby community college. By then a strong supporter of nuclear power, he began attending meetings of the Diablo Canyon Independent Safety Committee. This committee is the only one of its kind in the country, established by the state government in 1989 because of the various controversies surrounding the plant. It consists of three experts, appointed by the governor, who hold three lengthy meetings each year and issue an annual report with recommendations. "What I heard during those meetings was public comments from people who really didn't have science and engineering backgrounds," Gene told me. "They had very strong feelings about nuclear power."

In 2013, after the closure of SONGS, Gene and a few other like-minded advocates "saw the writing on the wall," as another member of his group put it to me. They were driven by frustration, even indignation: as the world grew more and more alarmed about climate change, people seemed to be willfully ignoring what seemed like the most obvious solution, a solution that was already at hand and had been providing reliable, low-carbon electricity for more than half a century. Instead of expanding the nation's nuclear fleet as rapidly as possible, the opposite was happening—the U.S. was shutting reactors down.

In the pro-nuclear community, Californians for Green Nuclear Power

(CGNP) stood out in several ways. One was the relatively advanced average age of their members. More than once, I heard other advocates refer to them as something along the lines of "these old guys." Most nuclear advocates were Gen Xers, like Shellenberger, or millennials, like Eric Meyer, who felt they had rediscovered a technology their elders had not appreciated. CGNP represented a demographic that was more associated with anti-nuclear activists, the OG environmentalist hippies. (With his long gray ponytail, Gene looked like an anti, which he realized and relished.) Whereas Gene had attended anti-nuclear marches before converting, most of the others, because of various biographical quirks, were pro-nuclear all along. Carl Wurtz, who served as the group's president for a time, grew up in Illinois, a state known for deriving a high percentage of its electricity from nuclear reactors. "In school in the sixties and seventies, we were taught that nuclear was the power of the future," he told me. "It was a completely different culture than California." CGNP was also a rare group—at least in California—that did not grow out of the Breakthrough Institute/Environmental Progress circle and initially had no connection with Shellenberger. When it came to saving Diablo Canyon, they were on the scene first.

They were also distinguished by their intensely local focus and their lower-profile, behind-the-scenes work. While Shellenberger gave TED Talks that garnered millions of views, Gene was talking to shoppers at the farmers' market and fielding listener calls on *Dave Congalton Hometown Radio*, a popular San Luis Obispo show; he was publishing op-eds and letters to the editor in the *San Luis Obispo Tribune*.[8] (Most, of course, focused on nuclear, but in one letter, he complained about recent changes to the newspaper's website: "No longer may I simply use my mouse to demarcate text and graphics to be archived. I must use a considerably more cumbersome process. Please restore the website design that you employed previously."[9]) He embodied the Beat poet Gary Snyder's advice to be "famous for fifteen miles"[10]: by 2019, people around town were starting to recognize him as "the guy in the headband."

He invariably showed up for hearings, meetings, and workshops related to Diablo Canyon. Whether it was the State Lands Commission, the Nuclear

Regulatory Commission, the Diablo Canyon Independent Safety Committee, or any of the myriad other commissions and agencies whose remit includes the plant, he was there, wearing his headband.

Mothers for Nuclear, the only other pro-nuclear group based in the San Luis Obispo area, engaged in some of those unglamorous activities, too. But what really set CGNP apart was that they were official intervenors with the California Public Utilities Commission. That meant that they could submit comments to the commission whenever a proceeding was underway.

In 2016, dozens of intervenors, including Mothers for Peace and Friends of the Earth, submitted comments about the plan to close Diablo Canyon. CGNP was the lone intervenor that argued for extending the plant's life beyond 2025. The proposal's "'clean energy vision,'" CGNP wrote in their comments, using scare quotes, "is one at odds with science, with public health, environmental health, California ratepayers, and the health of the California economy." They also wrote responses to the comments of the other intervenors. All of this added up to thousands of pages over the years. Heather and Kristin sometimes smirked at Gene's idiosyncrasies, but they ultimately expressed deep gratitude that he and his associates were doing this grunt work.

When the deal appeared to be done in 2018, Gene and his colleagues had already been trying to think creatively about how to save the plant. At their regular meetings—on the third Tuesday of every month at a Coffee Bean & Tea Leaf—they threw around ideas over cinnamon rolls and caffeinated beverages. Among their various ideas was a two-pronged strategy.

The first prong involved somehow getting state law to recognize the climate benefits of nuclear. A handful of other states had already implemented such laws, through mechanisms such as zero-emissions credits that could subsidize any low-carbon energy source. These made nuclear plants more competitive and enabled several financially struggling plants to survive.[11] Even some recent California legislation had nodded in this direction, specifically a 2018 law authored by Kevin de León, at the time a state senator and a progressive darling. (He would lose that status in 2022, while serving on the Los Angeles City Council, after he was caught on tape in a conversation that included racist and homophobic comments.) The law mandated that all electricity retail sales and electricity procured for state agencies be renewable *or zero-carbon* by 2045. While the law didn't specifically mention nuclear, the language certainly seemed to introduce some wiggle room. On its own, it wasn't enough to save Diablo Canyon—the timeline was too far out. But it suggested to Gene and other members of his group that the zeitgeist in the state legislature might be shifting to be more open to nuclear.

The second prong was more farfetched. It was an effort to find a buyer for the beleaguered plant—"a consortium who appreciates the value of Diablo Canyon," Gene told me.

By this time, in 2019, PG&E was again in deep legal and financial trouble. The company was facing charges of involuntary manslaughter for its equipment's role in sparking the Camp Fire, which obliterated the town of Paradise, California, killing at least 84 people. The utility's liability for that and other wildfires amounted to about $30 billion, leading them to declare bankruptcy for the second time on January 29, 2019.[12] Amid these crises, PG&E was not in a position to discuss reversing course on Diablo Canyon. Gene and his associates thought another entity might be interested in acquiring the plant. But this would become more plausible if the first prong of the plan worked out. A change in the law could boost Diablo Canyon's value as an asset and make it attractive to a buyer.

Of course, not just anyone can go shopping for a nuclear plant. It would have to be a major conglomerate with a great deal of nuclear experience. Gene told me he was in talks with interested parties, but declined to go into more detail.

CGNP's lawyer recommended that they hire a lobbyist. They initially balked, according to Carl Wurtz, because they thought of lobbyists as corrupt backroom dealers. (There were also limits on lobbying activities for 501(c)(3) organizations that they would need to obey.) But their lawyer cast the enterprise in a more benign light, convincing them that it was a matter of building relationships and explaining concepts to harried lawmakers who didn't have time to understand every issue in depth. Still, lobbyists were expensive. The group had some funds. As intervenors, they were eligible to apply for reimbursement from the state for time spent on their efforts, and they'd received significant payments that way; they had also received several generous private donations. But to defray the costs, they reached out to the other big player in their crusade: Michael Shellenberger. He agreed to chip in.[13]

The lobbyist focused his attention on Republican Assemblyman Jordan Cunningham, who represented San Luis Obispo. He was, like many Republicans, strongly pro-nuclear. But Cunningham was a California Republican; unlike some in his party, he acknowledged climate change and cited it as a reason for his support of nuclear. In August 2019, after discussions with the lobbyist, Cunningham filed a proposed constitutional amendment that would prohibit the legislature from discriminating against any form of zero-carbon energy, in effect reclassifying nuclear so that it would count toward the state's renewables targets.

When I spoke with Cunningham, he seemed to concede that it was a long shot. I wondered why he had proposed a constitutional amendment, which would face steeper odds than a bill. He said it was basically too late to introduce a bill at that point in the legislative calendar, and then implied that he didn't expect it to pass. "A lot of times this stuff is about driving a conversation, too," he told me. "It's not just, 'Hey, I got a bill signed that's going to name a freeway after somebody.'"

On August 29, 2019, Gene went into the studio to appear on *Dave Congalton Hometown Radio*. Congalton, a staple of the San Luis Obispo community, was a gruff but friendly presence, game to talk about anything from the most serious issues to minor indignities suffered by local residents; also on the show that day were the group Central Coast Vegans and a woman who had gotten a traffic ticket for the offense of activating her turn signal too soon.

The last time Gene had been on, Congalton had challenged him to find a local politician who would take up his cause. Now Gene was back to triumphantly discuss Cunningham's proposed amendment. He also mentioned the strategy of seeking a buyer, which he acknowledged was "challenging" in that "you have to be really qualified because obviously nuclear power plants need a lot of TLC."

Congalton was skeptical. "We also have a governor who's made it clear that he's not a fan of Diablo Canyon," he said. "Governor Newsom. So how do you change his mindset?"

"We play a deep game," Gene replied. "So we actually talked with his chief of staff when he was lieutenant governor, and we had a very favorable conversation, and we're keeping those lines of communication open."

"You think Gavin Newsom's going to change his mind?"

"I think it's a very real possibility, yes."

"What would be his motive?"

"Probably to get reelected." There was a pause, during which Congalton was presumably looking at Gene in disbelief. "Seriously," Gene went on, "this plant is so incredibly important to this state. In other words—you're old enough to remember Enron, right? The rolling blackouts and the traffic lights out and people worried about whether they'd be able to go to the hospital and get their operation done or not. That's actually what's in store if we don't get our ducks in a row."

Gene told the audience how to find CGNP's website: "Charlie Gulf November Papa dot org." He announced that the group was planning to offer some educational programming at the local library, because an informed citizenry would be more likely to make good decisions. And Congalton reminded his listeners that Gene would be in front of Mother's Tavern at the farmers' market that evening.

In the months to come, however, the efforts to find a buyer stalled, and Cunningham's amendment predictably didn't go anywhere. Meanwhile, at Diablo Canyon, a whole team was working on planning for the decommissioning. (Kristin Zaitz, to her chagrin, was on that team.) In anticipation of shutting down, the plant was deferring maintenance on some equipment. There was less and less time for the work that would be needed to reactivate the license renewal application. Trying to save Diablo Canyon seemed like a fool's errand, and Gene Nelson and the other Californians for Green Nuclear Power were looking more and more like crackpots.

CHAPTER FIFTEEN

THE INSIDE GAME

On January 27, 2016, Senator Cory Booker, wearing a dark suit and holding a microphone, paced back and forth in front of a small audience. "We are in a crisis, and more people need to wake up to that," he said. Booker, the junior senator from New Jersey, was a celebrity politician—a Stanford alum and a Rhodes Scholar who had returned to his home state of New Jersey and lived in the projects in Newark before becoming that city's mayor. He had been named by *Town and Country* one of the "Top 40 Bachelors" in 2013 and was known for his vegan diet. "When the planet is in peril, we are the nation that should be leading the way forward," he said. To reach "that big hairy audacious goal" of decarbonization, "it can't be done with wind and solar alone." Also necessary, he said, would be "new advanced nuclear energy."[1]

The occasion was the first Advanced Nuclear Summit and Showcase, sponsored by Third Way, the center-left think tank. Third Way had rented out the top floor of the Newseum, a journalism-themed museum, which occupied a seven-story building on Pennsylvania Avenue. A few days before the event was scheduled to take place, a blizzard had struck Washington. Josh Freed, Director of the Climate and Energy Program at Third Way, had worried that no one would show up. But the participants had braved the weather—another scheduled speaker was Senator Lisa Murkowski of Alaska, and her staff had assured Freed that a little snow wouldn't stop her—and now they were overlooking the National Gallery of Art and the National Mall, blanketed in white.

After the meeting at Ray Rothrock's house in Portola Valley in 2014, Freed, along with a handful of others, including Rothrock and Rachel Pritzker, had begun scoping out the political mood in D.C. They weren't sure what to expect. The long-standing assumption was that Democrats were firmly anti-nuclear. There were some exceptions, such as the Illinois Democrats who'd defended the Integral Fast Reactor. But in general, as Peter Teague, who had worked on the Hill, put it, "If you were an environmentalist, if you were a progressive, liberal, there was just no question but that you were anti-nuclear. It was part of the DNA."

Counterintuitively, though, Pritzker, who had been deeply involved in progressive politics, saw nuclear as a possible way of cutting through polarization. The key was to focus on advanced nuclear, emphasizing novelty and innovation. Democratic officeholders had bad associations with existing nuclear, but maybe advanced nuclear could present an opportunity to, in effect, rebrand. Also, circumstances had changed, even since the '90s, when John Kerry and Bill Clinton had killed the Integral Fast Reactor. Now, 20 years later, climate change had moved from journal articles to the daily news and had become a much more salient issue for Democrats. And authoritative sources such as the UN Intergovernmental Panel on Climate Change (IPCC) were saying that nuclear would be necessary—or at least very helpful—in meeting climate goals.[2]

When these advocates first started meeting with lawmakers, they encountered some resistance. In particular, Dianne Feinstein, the late, long-serving senator from California, had made it clear that the lack of resolution to the waste was a major issue for her. But the lawmakers and their staffs didn't completely shut the advocates down. (Another benefit of focusing on advanced nuclear was that some of the designs were intended to run on spent fuel—so they could in theory present a solution to the problem that so concerned Feinstein and some other Democrats.)

As Freed and his colleagues at Third Way started to survey the landscape, they also discovered something surprising: that a number of advanced nuclear start-ups had already been launched. In 2015, the think tank developed a map of all of the companies pursuing advanced nuclear, including fusion. They identified almost 50, which was far more than they had expected. They found start-ups working on a dizzying variety of technologies: molten salt reactors, liquid metal–cooled fast reactors, high-temperature gas reactors, pebble bed modular reactors, small modular reactors, and something called a "nuclear battery."[3] On the map, dots representing start-ups were scattered all around the country, albeit more concentrated in California and the Northeast. Equally striking, these companies were flush with private funding: they had an estimated total of more than $1.3 billion, all from private investors. Historically, the nuclear industry had been heavily reliant on government funding; private investment, at least in R&D, was fairly unprecedented.[4] To the analysts, this interest was a promising sign of nuclear's prospects.

Still, these companies were facing formidable hurdles to getting off the ground. In the history of the Nuclear Regulatory Commission, the agency (like its predecessor, the Atomic Energy Commission) had granted licenses to only a single kind of facility: the light-water reactor. The NRC's approach would need to be reformed in order to handle applications from a range of companies pursuing a variety of technologies. Another factor was that start-ups needed to be able to test and refine their designs, but that wasn't something you could just do in your garage. For that, they would require government support—specifically, they would need to be able to take advantage of the immense resources at the Department of Energy's National Labs.

In 2015, Josh Freed, along with Pritzker, Rothrock, and Ross Koningstein (another funder) met with Ernest Moniz, Obama's secretary of energy at the time. They went to his office in the Forrestal Building, a brutalist building a few blocks east of the Mall. "We literally sat in his office and showed him this map," Freed told me, "and he handed it to his then–deputy chief of staff and said, you know, let's see if this is right. And it was."

Obama was already receptive to nuclear power. After Fukushima, his administration had affirmed that nuclear would continue to play a key role in the country's energy system, pointing out that reactors provided 20 percent of the nation's electricity and more than half of its low-carbon electricity. It soon became clear that he also wanted to nurture advanced nuclear.

The first win for the advocates came in November 2015, when the Obama administration announced the formation of the Gateway for Accelerated Innovation in Nuclear (GAIN) initiative. One goal was to link the new start-ups with the National Labs, to enable them to test their novel designs in a way they couldn't do anywhere else. The labs had long been a place to carry out research for the sake of research, but there had not been an obvious pathway to commercialization. The purpose of GAIN would be to bridge that gap, helping these start-ups actually become commercially viable.

Meanwhile, Republicans, as we've seen, had generally never had a problem with nuclear. The reasons for their support were not typically climate-related; they were more likely to be on the grounds of energy security (and as a bonus, owning the libs). But nuclear also had the potential to open up an avenue for conservatives to act on climate by touting a solution they already liked. Another D.C.-based nonprofit, ClearPath, approached climate from a conservative perspective, and they were also talking with lawmakers about nuclear during this period. A small group of Republicans, including Lisa Murkowski of Alaska and Mike Crapo of Idaho, began working with Democratic senators, including Booker and Sheldon Whitehouse of Rhode Island.

At the Third Way event in January 2016, the momentum for bipartisan legislation was building as the speakers emphasized their various reasons for embracing nuclear. "Now, I come to this crisis from a very different way than I imagine most of the people here," said Booker. He brought up asthma, which can be both caused and exacerbated by air pollution. "I go home every week to Newark, New Jersey, and I talk to parents who have kids that are missing school," he said, because of asthma. "It is killing the potential of my children. . . . The particulate matter is pouring into New Jersey's airspace . . . because of things like burning coal."

Rachel Pritzker introduced Senator Murkowski, who came up to the lectern in a beige sweater and dangly earrings. She thanked Third Way for "challenging us as policymakers, as lawmakers, to think outside of the partisan box." She went on, "When you want to talk about reliable and clean, the first place that you should be looking is nuclear. In my mind it's as simple as that." In Alaska, she said, small modular reactors could be "a game changer for us, in some of our remote communities, our small communities. . . . I see advanced reactors as the next chapter of America's leadership on nuclear technologies."

The two parties had different reasons for supporting nuclear, but it didn't matter as long as they could come together on legislation to do so. Murkowski announced that that very day, they would be taking up a bipartisan energy bill in the Senate that would include provisions on advanced nuclear.

THE POLITICS SURROUNDING NUCLEAR ENERGY HAD ALWAYS BEEN COMPLEX and full of paradox. The inherent properties of the energy source didn't line up neatly with a particular political party or system. Certain characteristics—the fact that nuclear plants were large infrastructure projects that required significant government support—fit in well with a more socialistic system. Indeed, the Soviet Union had built many nuclear plants. Today, several of the countries known for their nuclear fleets—Sweden, Finland—are also viewed by many on the left as close to social democratic utopias.

As a perceptive Breakthrough Institute report, authored by Ted Nordhaus and Kenton de Kirby, put it, "Despite the fact that nuclear is a technology that has historically required the heavy hand of government, it has been rejected by those who believe in active government but embraced by libertarians and conservatives. It represents a public good largely developed and operated by the government but has been widely condemned as a corporate conspiracy." The report went on, "Clearly, nuclear sits in a liminal space . . . in the public mind, it is neither fossil fuel nor clean energy . . . public attitudes about nuclear energy might be quite open to revision."[5] They also noted that a new generation of small modular reactors might present an opportunity to reshape these opinions, because such reactors would fit into the model—small-scale, distributed energy—that progressives and environmentalists tended to favor.

In 2018, Congress passed, and President Trump signed, the Nuclear Energy Innovation Capabilities Act (NEICA), which also sought to facilitate advanced reactor development. What was particularly notable was how the bill passed. The senate passed it by voice vote, in which votes are not recorded, because no members raised serious concerns about it; the House passed it under suspension of the rules, for the same reason.[6] The same year, both the House and Senate voted overwhelmingly to approve the Nuclear Energy Innovation and Modernization Act (NEIMA), which Trump signed in early 2019. This law was directed at the NRC, intended to ensure that it could accommodate the new kinds of reactors that were in the pipeline. Two months later, Senator Murkowski introduced yet another relevant bill: the Nuclear Energy Leadership Act (NELA), which directed the secretary of energy to establish goals related to advanced nuclear, including securing a fuel source, and to set up a demonstration reactor. This bill was also bipartisan, with 9 Democratic cosponsors, including Booker and Whitehouse, and 13 Republican cosponsors. It passed in July 2020.

Nuclear energy had gone from liberal bogeyman to one of the very few points of agreement between the parties. And not only did a core bipartisan group of senators diligently work to pave the way for advanced nuclear; equally striking was the absence of effective organized opposition. As it turned out, even most of the big green groups—NRDC, Environmental Defense Fund, the Nature Conservancy—had, to varying degrees, come around to the proposition that nuclear would be needed for climate reasons. In other words, it seems that Rachel Pritzker was onto something. During a period defined by extreme partisan rancor, when the parties saw each other as increasingly monstrous and couldn't even always manage to keep the government running, this was truly remarkable.

This is not to say that all opposition to nuclear had disappeared. In 2018, there was vigorous debate about whether it should be included in the Green New Deal—the left's new plan to address climate change and create jobs through a massive build-out of clean-energy infrastructure. (It bore more than a passing resemblance to the New Apollo Project, the initiative Nordhaus and Shellenberger had been part of in the early 2000s.) Just after the 2018 midterm elections, the Sunrise Movement, the youth-led climate justice organization, occupied Democratic leader Nancy Pelosi's office, holding yellow signs that read, "What's your plan?" The newly elected star congresswoman Alexandria Ocasio-Cortez turned up to show her support. On the same day, AOC released a draft resolution outlining what the Green New Deal should seek to accomplish. The first objective was "100% of national power generation from renewable sources." In January 2019, an open letter signed by hundreds of environmental groups, addressed to members of Congress, demanded 100 percent renewables by 2035, and stipulated that "any definition of renewable energy must also exclude all combustion-based power generation, nuclear, biomass energy, large scale hydro and waste-to-energy technologies." The signatories included Friends of the Earth, Greenpeace, 350.org (the group Bill McKibben cofounded but no longer ran), and many smaller local groups—though it was perhaps more interesting to observe who was missing, such as the Environmental Defense Fund and the Sierra Club.[7]

Yet, despite the initial language of the resolution, the single individual most closely associated with the Green New Deal, Representative Ocasio-Cortez, demonstrated increasing openness to nuclear over time. In an interview in May 2019, she said that her resolution "leaves the door open on nuclear so that we can have that conversation."[8] A climate platform that she subsequently coauthored for the Democratic Party with John Kerry—a quarter-century after his campaign to terminate the Integral Fast Reactor—called for "cost-effective pathways" to develop advanced reactors.[9]

None of the new nuclear-friendly laws got a lot of media attention at the time. And, from the perspective of pro-nuclear advocates, that may have been a good thing. Dating back to the movement's early days, there had been debate about how much to focus on public opinion campaigns versus what Freed called "the inside game." In part, this was a matter of allocating finite resources, but it was not only that. The advocates also thought there could be benefits to a more behind-the-scenes approach that did not make the issue salient. One factor was that some segments of the public might be harder to persuade than lawmakers, and a higher profile might translate into more negative attention. Another was that in a hyper-polarized electorate, bipartisanship was not always viewed in a positive light. If certain voters noticed that their representatives were reaching across the aisle, the collaboration could turn into a liability.

As support for advanced nuclear grew, so did recognition of the value of the existing fleet. In October 2018, a report about the risks of early closures was issued by a surprising source: the Union of Concerned Scientists. The group had been founded in the late '60s in part to raise alarms about the hazards of nuclear plants. Now the cause of their concern was that too many plants might shut down. In a publication titled "The Nuclear Power Dilemma," they noted that about a third of the country's nuclear power plants were at risk of premature closure or scheduled to retire. The report warned that "closing the at-risk plants early could result in a cumulative 4 to 6 percent increase in US power sector carbon emissions by 2035."[10] (One of UCS's longtime leaders told me that the group had never been strictly anti-nuclear; they had merely sought to improve safety.)

In the 2020 presidential campaign, Biden was solidly supportive of nuclear, which caused little controversy. For the first time in decades, the official Democratic Party platform endorsed nuclear: "Recognizing the urgent need to decarbonize the power sector, our technology-neutral approach is inclusive of all zero-carbon technologies, including hydroelectric power, geothermal, existing and advanced nuclear, and carbon capture and storage," it read.[11] The Bipartisan Infrastructure Law, passed in November 2021, included $2.5 billion to continue supporting advanced reactor development. It also included a provision that did not go unnoticed back in California: $6 billion in funding for the Civil Nuclear Credit Program, which was intended to preserve existing reactors that were threatened with closure.

CHAPTER SIXTEEN

"A HERETIC AMONG HERETICS"

WHILE THE PRO-NUCLEAR MOVEMENT DID NOT HAVE A SINGLE CLEAR leader, in the late 2010s, Michael Shellenberger, with his TED Talks and Twitter following, became the most public face of the movement. For defending nuclear in debates on university campuses or on radio segments, he was often the one tapped. He was a polished speaker, he was not shy of attention, and his combative style made him more memorable than others who may have been more diplomatic.

This, increasingly, posed something of a conundrum for the movement. While he certainly drew some people in, he turned other people off, and the net effect was hard to calculate. He had long been caustic in a way that made some allies uncomfortable. He claimed to be "a lifelong environmentalist, progressive, and climate activist,"[1] but ever since the publication of "The Death of Environmentalism," he'd routinely excoriated environmentalists, progressives, and climate activists. Representative headlines from his *Forbes* column included "Why Earth Overshoot Day and The Ecological Footprint Are Pseudoscientific Nonsense"[2] and "California Has Always Had

Fires, Environmental Alarmism Makes Them Worse Than Necessary."[3] He defended plastic[4] and attacked Greta Thunberg.[5] He began appearing on Tucker Carlson's show on Fox News.

And he became more controversial with each passing year. In 2020 he published a book titled *Apocalypse Never: Why Environmental Alarmism Hurts Us All*, arguing that the threat of climate change had been overhyped. In 2021, he veered into new territory, publishing a book called *San Fransicko: Why Progressives Ruin Cities*, in which he argued that California's tolerant policies on drugs and encampments were responsible for the homelessness crisis. In 2022, he blurbed a book called *Fossil Future: Why Global Human Flourishing Requires More Oil, Coal, and Natural Gas—Not Less*. All of this raised questions about who he was trying to reach (wasn't the goal to convince progressives to embrace nuclear, since conservatives were already for it?), how focused he was on the cause (homelessness was important, but wasn't it a bit off topic?), and what his strategy was (if neither climate change nor fossil fuels were so bad, then why did we need nuclear, again?).

Anti-nuclear activists suspected Shellenberger of being some sort of Svengali. As the pro-nuclear movement emerged, opponents were convinced that it must be engineered by Shellenberger, whom they suspected, in turn, of being financed by industry. When I spoke with anti-nuclear activists about Mothers for Nuclear, they implied that the women were Shellenberger's puppets. While it was clear to me that Kristin and Heather were too assertive and sincere to be puppets, it was true that without Shellenberger's early encouragement and support, their group would not exist, nor would Eric Meyer's nonprofit, Generation Atomic.

By 2020, several new pro-nuclear groups had come onto the scene—Stand Up for Nuclear, Campaign for a Green Nuclear Deal, and Radiant Energy Group. They were, I would learn, all founded by former employees of Shellenberger's at Environmental Progress, and would also likely not exist without him.[6] (The founder of Campaign for a Green Nuclear Deal was Madison Hilly, at the time Madison Czerwinski, the young woman who was photographed pressing her baby bump against a waste canister.) So the notion that he had a hand in a lot of the pro-nuclear activity, sometimes in ways that were not obvious, was on the mark. And although, by all accounts, he did not take industry money—"Never have, never will," he liked to say—he did broker arrangements to secure industry funding for one of these spin-off groups.

His connection to these organizations was not exactly a secret, but it wasn't always out in the open, either. On November 9, 2020, Shellenberger published a *Forbes* piece in which he referred to "Madison Czerwinski, founder of a new group, Campaign for a Green Nuclear Deal," neglecting to mention that she was his former employee. In the same piece, he quoted Mark Nelson of Radiant Energy Group, also without disclosing their connection.[7] To the casual observer, the pro-nuclear movement might seem bigger than it was.

Most of Shellenberger's peers—former colleagues at the Breakthrough Institute and that larger circle—were dismayed by his trajectory. Descriptions I heard included "loud," a "narcissist," and "an anti-liberal troll." "I don't know what his game plan is," one baffled former associate told me.[8] Some of those who saw him as a mentor, though, tended to be more sympathetic, acknowledging that he was inflammatory but discerning strategy in his choices, or taking an attitude along the lines of, "Oh, that's just Michael being Michael."

Among these protégées was a young woman named Paris Ortiz-Wines. She graduated from UC Santa Cruz in 2017 and started applying for jobs with environmental organizations: big groups like the Sierra Club and NRDC, but also smaller nonprofits she came across in her search. One of them was Environmental Progress. As part of the application process, Shellenberger assigned her to watch his own TED Talk, "Why I changed my mind about nuclear power." Paris had never given much thought to the subject, but she was an easy convert. Whether because of her youth and resulting lack of nuclear-related baggage, or the persuasiveness of Shellenberger's talk, or some other idiosyncratic factor, her reaction was simple: "Yeah, I'm on board."

She started working at Environmental Progress in January 2018 as Shellenberger's executive assistant. The office was on Telegraph Avenue, the main thoroughfare near the campus of UC Berkeley, so the area was filled with students. Just down the block was Buffalo Exchange, a vintage clothing store, where Paris would sometimes go shopping after work. There were only two other employees: Madison Czerwinski, the executive vice president, and Mark Nelson, the senior analyst. By the end of 2020, Shellenberger was shifting his focus to homelessness and losing funding because of concerns over the direction of his work. That's when his employees began going off on their own. Nelson started Radiant Energy Group, a consulting firm with both industry and nonprofit clients. Paris joined him and, as a volunteer, kept up a project she'd worked on at Environmental Progress called Stand Up for Nuclear. She spent her days on calls with advocates all over the world, and also worked with people closer to home on saving Diablo Canyon.

If Heather and Kristin represented one strand of unexpected nuclear advocacy—hippie-ish, tree hugger moms—Paris seemed to epitomize another demographic that had not historically been associated with the nuclear industry: she was a young woman from a Mexican American family, and she was socially progressive. Her Twitter bio not only included her pronouns but listed them in English and Spanish ("she/ella"). She was a witty, charismatic presence on social media. Her pinned tweet, which had gone viral, was a photo of her in a hard hat smiling and standing on a concrete pad next to a large vault filled with spent nuclear fuel. The caption reads: "Everyone: What about the waste?? Me: What about it?" (This was a few years before Hilly would one-up her with the pregnant-belly shot.)

In March of 2022, I met Paris at a café in Berkeley, where we sat at an outdoor table in the sun. She told me she saw her time at Environmental Progress as her "second form of education," supplementing and sometimes correcting what she learned at UCSC—that is, the old-school, small-is-beautiful environmentalism that saw humans as the problem and saw the problem as apocalyptically dire. "After college, at every party, I was like, 'This is what we need to do, people need to stop eating meat, the world is just so messed up.'" At EP, she said, she learned to see environmental issues in a more measured way. It wasn't all doom and gloom; there were technological solutions to our problems. "There are some beautiful things that humans have created that can help us," she said. Foremost among them, of course, was nuclear energy.

She had not always agreed with her former boss, but she said there was "no bad blood" between them. "What I would say with Michael," she went on, "he's not everybody's cup of tea. He has a very strong personality." Then she offered what I considered a generous theory of Michael Shellenberger. She proposed that his strategy is to be the first one to make unpopular statements—pointing out the shortcomings of renewables, or "commenting on how eco-anxiety is really taking over our youth." And then, a year later, others can make similar comments in a more diplomatic way: maybe wind and solar do have some limitations; maybe some of the more extreme panic about climate change—the world is ending in 12 years!—was exaggerated or counterproductive. "So I would say he's a trendsetter but not well liked."

THE DAY AFTER MY COFFEE WITH PARIS, I HAD THE CHANCE TO POSE MY questions to Shellenberger himself. I drove from my Airbnb to Shellenberger's house, ascending from the flat area near downtown Berkeley, where modest homes sat on small lots and yards tended to be overgrown with wildflowers, to the Berkeley Hills, a beautiful area with live oaks, bay laurels, and eucalyptus trees, where the homes were larger and farther apart, and the gardens more manicured. Shellenberger and I had talked on the phone previously, but I was feeling slightly nervous to meet with someone who seemed to make a point of being intimidating and combative. At the same time, I knew that people who fostered a cult of personality, as he seemed to do, could be charming, quick to flatter, so I wasn't sure what to expect.

I parked near the address he had given me and walked over to his house, where two men were doing some sort of work out front. He'd instructed me to meet him in his home office in the back. I entered a gate to the right of the driveway and found a staircase leading down to the backyard. In an email, he'd written, "Please be careful walking down as it can be slippery especially if it rains." I kept descending some stone steps into the backyard, then realized I had gone too far, turned around, and saw him looking out the window of his office.

When I entered the room, he invited me to sit on a plush sofa as he sat on his desk chair. His desk faced a large window overlooking his backyard, with a view of the bay in the distance. The study was airy, quiet, and filled with light. To my right was a bookcase with shelves of books about nuclear weapons. I requested permission to record our conversation. He asked if we could speak on background first, and I agreed. I found him to be neither aggressive nor ingratiating. He struck me as very serious, almost solemn, and guarded, carefully choosing his words. He did not seem eager to please, but at one point, he did seem eager to be helpful. As I was fumbling around in my bag for a pencil, he hastened to find one of his own, examined it, sharpened it with an electric sharpener, and handed it to me.

I asked him to tell me about his background—how he came to be radicalized as a high school student, when, as he mentioned often, he'd gone to Central America to try to help the Sandinistas, the left-wing Nicaraguan political party that was in a civil war against the U.S.-backed Contras.

"I think I had a messianic complex resulting from my parents' divorce. And some amount of my father projecting aspirations of a kind of world-saving fantasy onto me, at a young age," he said. "I was disagreeable but also wanting to be liked."

Paris's comment fresh on my mind, I asked, "Do you still want to be liked?"

"Yes," he said, after a slight pause, and then slowly, "but other goals are more important to me now."

As he tells it, he does not fear controversy but also doesn't court it for its own sake. "Controversy is part of social change. And so is conflict. So I don't mind at all that it's controversial. But I do bristle at the suggestion that it is in some ways gratuitously polarizing or off-putting." He went on, "I don't write about or speak about or work on things where the majority is with me. There's no point. Life is short. Trump was chaos embodied. Who needed me to say that?"

"What about the idea of building bridges and finding common ground?" I asked.

"I think I've done an exceptional job of it and have an exceptional record of doing that," said the author of a book subtitled *Why Progressives Ruin Cities*. "I do resent when people accuse me of not doing that." As an example of his bridge building, he cited his TED Talks. "They always start where people are at. . . . I think the reason those TED Talks have been so successful, with millions of views—I get letters every week from people that watched them and say they changed their mind. It's because I do that. I take it extremely seriously."

About an hour into our conversation, he got a notification on his phone. After he put it back down, before the screen went black again, I saw a familiar shade of cobalt blue. It was the color of the recording app I used. I paused for a moment. "Oh, are you recording it, too?" I asked. "Yeah," he replied. "Is that okay?" If he was flustered, he hid it well. He had not been using his phone at any other point since I'd entered the room, so I can only assume that he'd been recording the entire time. In other words, when he asked me to wait before starting to record him, he was already recording me.[9] I quickly scanned what I'd said so far and was relieved that I couldn't recall anything particularly embarrassing, beyond my usual awkwardness as an interviewer.

It became clear as we spoke that he saw himself as something of a martyr to nuclear energy. He said he had "sacrificed significantly for this technology," ending his friendships with Nordhaus and Peter Teague and leaving the Breakthrough Institute. "I also quit drinking for this technology. I had a drinking problem. I stopped drinking in part so I could read at night, because I needed to read every major book that's ever been written on nuclear weapons." That was because he was now writing a book about nuclear. He gestured toward the bookshelves to my right. "Those are just the weapons books. All the books on energy are upstairs. And I couldn't read while drinking at night."

But he was a martyr for his particular vision of nuclear energy. After all, Nordhaus and the Breakthrough Institute were pro-nuclear as well.

Shellenberger differed from them in two substantive ways. The first was that he was so critical of renewables. He dwelled on the downsides. For example, he would point out that since solar panels only last about 25 years and there are currently few options for recycling, they have begun to pile up in landfills; some contain toxic heavy metals such as lead and cadmium.[10] "There was definitely a moment when I was with a bunch of our allies and I was kind of like, 'If you have nuclear, why do you need renewables, which cause all these problems?'" he said. "My colleagues were like, 'You can't go there' and whatever, and I was like, 'Well, why?' And it was, 'Well, because people don't like nuclear, and renewables still have some role or something something something.' Nobody had a good reason why."

I asked if he thought renewables had any role to play.

"We have solar panels in our backyard," he said, gesturing out the window. "Solar panels are a niche that are useful for some off-grid and backyard applications. They're terrible at scale. Devastating." Then he uttered a line that was more familiar coming from his anti-nuclear adversaries: "There's no solution to the waste."

The other dispute between Shellenberger and the majority of pro-nuclear advocates involved advanced nuclear. After all, there were a lot of ways to design nuclear reactors, and the recent record in the U.S. of building the old style was discouraging, with several units going billions of dollars over budget and years past their scheduled opening, if not getting outright canceled.[11] So, most advocates thought, why not push the next generation of innovative designs?

Shellenberger's take was that the existing nuclear fleet was great, something to be proud of. Like anti-nuclear activists, in fact, he thought notions of advanced nuclear were a fantasy. "They talk about small modular reactors, they talk about these gas-cooled or other kinds of reactors, but they don't happen," he said. He thought that hyping them tarnished the reputation of the existing reactors, implying that the latter were second-rate and, especially, unsafe. "We know that what makes nuclear cheap, safe, reliable is experience. So it's experienced people building the designs, it's experienced people operating and regulating the designs, that's what matters." He added,

"This thing of making them smaller is idiotic," undermining the economy of scale that is one of the advantages of a large nuclear plant.

"I'm a heretic among heretics," he said, not without pride—a line I would hear him repeat publicly in the future.

BEFORE MY TRIP TO THE BAY AREA, I'D CHECKED SHELLENBERGER'S BOOK *Apocalypse Never* out of the library and started reading it, intending to finish it before our conversation. But I'd put it down after the first couple of chapters.

I was receptive to the argument that climate change was less apocalyptic than I feared. After all, that would be great! What put me off about the book was his approach to making this case. In the introduction, for example, he points to a debunked study about the effects of carbon dioxide on coral reef fish behavior. It turned out that the researcher had fabricated data. He cites the debunked study along with several other data points showing that climate trends are more positive than we are told. "Does any of that really sound like the end of the world?"[12] One bad study, he implied, meant that there was no need to worry about coral reefs. *Does he think his readers are dumb?* I wondered. Then I realized: most of his readers are likely people who already think climate change is overhyped, and we are always primed to gloss over logical fallacies to support our existing worldview.

In the larger story of the pro-nuclear movement, *Apocalypse Never* and its reception were telling. Some of Shellenberger's former allies criticized aspects of the book. (Alex Trembath called him out for "nuclear fetishism" and for downplaying biodiversity losses.)[13] But the larger point—a critique of "alarmism" or "catastrophism"—was common, though not universal, among pro-nuclear advocates.

I found this puzzling at first. The only reason I had become open to nuclear energy was that I was so alarmed about climate change. Nuclear seemed scary, and for a lot of people, the only reason to consider it was that it could help prevent something even scarier. To me, it seemed like nuclear advocacy and climate alarm would, or should, go hand in hand.

What I eventually realized was that there was a certain coherence to their position. They saw nuclear panic and climate panic as stemming from the same misguided worldview—a quasi-religious attitude toward nature, seeing it as something sacred that humans tamper with at their peril. A 2015 document titled "An Ecomodernist Manifesto," published not long before Shellenberger's split with the Breakthrough Institute, was an attempt to crystalize an alternative philosophy; its 18 signatories included Nordhaus, Shellenberger, Peter Teague, Stewart Brand, Mark Lynas, Rachel Pritzker, and Robert Stone. In stark contrast to the dominant environmentalist narrative, their vision of the future was rosy. "As scholars, scientists, campaigners, and citizens, we write with the conviction that knowledge and technology, applied with wisdom, might allow for a good, or even great, Anthropocene," they wrote, invoking a term that had been coined to describe this era of human influence on the planet. "A good Anthropocene demands that humans use their growing social, economic, and technological powers to make life better for people, stabilize the climate, and protect the natural world." Building on Brand's ideas in *Whole Earth Discipline*, they proposed that "[i]ntensifying many human activities—particularly farming, energy extraction, forestry, and settlement—so that they use less land and interfere less with the natural world is the key to decoupling human development from environmental impacts."[14]

The idea that we shouldn't freak out about climate change was so foreign to me that it took some time for me to grasp what they were saying. But I came to understand that, as opposed to the classic environmentalist worldview, ecomodernists emphasize that nature has never been static and that significant human influence on the natural world dates back at least to the beginning of agriculture, placing climate change on a continuum with those other changes. They point out that huge energy inputs have yielded the benefits of modern society, including human rights advances. For example, although the relationship between the industrial revolution and slavery was complex, over the long term, the availability of abundant energy—literally defined as "the ability to do work"—was likely one factor that contributed to slavery's decline.

Another key point was that, even as climate change has made weather extremes more likely, intensifying floods and hurricanes and droughts, humans have become far more resilient to these events. Amazingly, deaths from these disasters have declined dramatically in recent decades. According to the respected site Our World in Data, which collates studies, in 1920 there were 523,892 deaths from natural disasters. In 2000, there were 78,130; in 2020, there were 41,046.[15] (And this despite a much larger population.) The reason is development: that as human societies develop, they build more robust structures for shelter; they build roads to enable evacuation; they learn meteorology to forecast weather; they establish communication systems to send warning signals; and they have hospitals to save the wounded. So, even as weather events have increased in severity, the death toll has continued to decline.

The challenge is to "decouple" energy use from greenhouse gas emissions and other environmental impacts, with the goal of stabilizing the climate and sparing land. Of course, one important way to do that, in this worldview, is to build lots of nuclear power plants.

As someone who had always been a pretty hard-core environmentalist, I had mixed feelings about this alternative philosophy. I thought it had the potential to be dangerous, instilling complacency about what I considered an ongoing cataclysm. It seemed too sanguine about the real possibility of tipping points that could lead to extreme disruption; it also struck me as too dismissive of all the casualties other than deaths, like the loss of seasons and stability. At the same time, I genuinely appreciated that it helped me see the situation in a more nuanced way—not just "Everything sucks and we're doomed," which was how I had been feeling for a long time.

I eventually finished reading *Apocalypse Never*, and I thought it included some valid points. But these were often undermined by the style of argument I'd noticed at the beginning. A group of scientists closely reviewed a viral article Shellenberger wrote to promote the book, and they concluded that it "mixes accurate and inaccurate claims in support of a misleading and overly simplistic argumentation about climate change."[16] Notably, one of the reviewers was MIT atmospheric scientist Kerry Emmanuel, who had served as a science advisor to Environmental Progress before leaving in frustration.[17]

In Shellenberger's office, I asked him about these criticisms. He said that those scientists were "activists" who wanted to "censor" him—even though one had been a supporter until recently. "Outrageous," he said. "Totally wrong on all of the facts."

So, although he seemed to feel unfairly penalized for expressing his views, he was also taking something of a victory lap, as he saw his former employees out there in the world and public opinion changing. "The pro-nuclear movement, it was just an idea, it was just a kind of . . . it was just bullshit. We were just bullshitting. It was classic fake it until you make it. But now there's genuinely a pro-nuclear movement. There's people in the pro-nuclear movement who I don't know. There's people in the pro-nuclear movement who *don't like me.*" That contingent of the movement would grow, as, after my visit, he would continue on his path toward becoming a full-fledged culture warrior and conspiracy theorist, staking out right-wing positions on transgender issues and disinformation research, and even wading into debates over UFOs.

And although at times he had arguably played down his influence, making the pro-nuclear movement seem bigger than it was, now he wanted to make sure that his central role in the movement was recognized. "I want everybody to get the credit they deserve, and I just don't—I don't want to overshadow anybody but I also don't want to be, like, portrayed as somehow outside of the movement that I consider myself as having created," he said. "Does that make sense?"

CHAPTER SEVENTEEN

NUCLEAR SPRING

In the central African country of Gabon, there is a region known as Oklo, where uranium deposits are rich and abundant. Evidence suggests that about two billion years ago, a random neutron split a U-235 atom in the ore. The neutrons from the split nucleus went on to split other atoms, in a self-sustaining chain reaction. Groundwater acted as a moderator, and the reaction continued, on and off, for several hundred thousand years. During this time, according to one scientist, the reaction generated, on average, enough energy to power a few dozen toasters.[1] It petered out as the fissile isotopes diminished, and as "nuclear poisons" built up that got in the way of further fissioning—exactly what happens in a nuclear reactor.

Thousands of millennia after the reaction ceased, French colonists found the uranium deposits and began exploiting them for their nation's civilian reactor program. In 1972, a worker noticed something unusual about the uranium ore: it contained a slightly lower concentration of U-235 than expected. Later, other elements—fission products such as lanthanum, cerium, and xenon—were discovered at the site. Essentially, the uranium seemed as if it had already been in a nuclear reactor. Scientists concluded that the fission had occurred spontaneously.[2]

The discovery is well known in nuclear circles. In 2013, when Jake DeWitte and Caroline Cochran, nuclear engineering grad students at MIT, founded an advanced nuclear start-up, they decided to borrow the name Oklo for their company. The appeal is clear: the story implies that nuclear energy is not the egregious violation of nature it had always been assumed to be; it is, on the contrary, a natural phenomenon. The name also sounds catchy, contemporary. The older nuclear industry stalwarts, which were not exclusively nuclear but rather energy companies and utilities, had names like Westinghouse Electric Corporation and Dominion Energy. The name Oklo signals the company's singular focus on nuclear and its youthful brand-savviness. This is what the next generation of nuclear wants to be.

One gray day in March of 2022, I visited the Oklo headquarters, in Santa Clara, California, a city in Silicon Valley. The offices have a tech-ish aesthetic—vast and open-plan, with smooth concrete floors and geometric, sculptural art on the walls. When I entered, I was greeted by Jake and Caroline, who were now married and in their 30s. Jake, the CEO, has a boyish, quick-thinking energy; Caroline, the COO, is friendly in a low-key way. They welcomed me into a conference area where we sat around a large oval table.

When I sat down, I reached into my bag and pulled out *Plentiful Energy*, the book about the Integral Fast Reactor, coauthored by the project's leaders, Charles Till and Yoon Chang. I'd brought it because I'd read that Oklo was building on that technology. The book is oversized with a bright blue cover, and Jake recognized it immediately. "We recommend it to all of our employees," he said. To Jake, it was tragic that the program had been killed. He saw Oklo in part as a tribute to the work of Till and Chang and their colleagues, ensuring that it wasn't for naught. He called them "extraordinary uncelebrated heroes in terms of climate."

Days before my visit, Russia had invaded Ukraine, and some of the major geopolitical repercussions already under discussion had to do with energy. Much of Europe, especially Germany, was dependent on imported Russian natural gas. Greater adoption of nuclear could be a way to alleviate that dependence. At the same time, the day before my visit, the Zaporizhzhia Nuclear Power Station—the largest nuclear plant on the continent, with six reactors—had become a war zone, raising alarms about an aspect of nuclear risk that few had even previously contemplated. Further complicating matters, Russia was also a major supplier of enriched uranium, needed for nuclear fuel, and the U.S. relied on Russia for almost a quarter of its uranium enrichment[3] —not an ideal position to be in.

One advantage of the reactor that Oklo is designing, called the Aurora, is that it would not require fuel from Russia because it is intended to run on recycled waste. The plan is to build the first prototype at Idaho National Lab, where work on the Integral Fast Reactor also took place. And, in a nice karmic twist, the Department of Energy awarded Oklo the right to use reprocessed used fuel from the storied reactor that served as their inspiration.[4]

Jake pulled up a PowerPoint presentation on his laptop, projecting it onto a large screen on the wall, going into what Caroline called his "professor mode," and explained the basics to me. The image they have created of the Aurora makes it look like a gleaming, high-tech version of an A-frame cabin. At the time of my visit, they were planning to build "micro-reactors" that produce only about 1.5 megawatts (compared with the 2,250 megawatts from Diablo Canyon). That's enough to power a small village—and among their first customers could be remote communities in Alaska, like the ones Murkowski invoked at the Third Way summit, which currently rely on imported diesel fuel. The Aurora could also power factories or data centers. Part of the motive for this smaller scale is that it will require less up-front capital investment. The promise of small modular reactors is that many identical parts could be prefabricated in factories. So, before long, the thinking goes, we would see learning curves and cost reductions of the kind that we've seen with renewables in recent years.

Like the IFR, it will incorporate passive safety features, meaning that it should be physically incapable of a meltdown. But, unlike the IFR, they are not planning to "breed" fuel—that is, create more fissile material than is consumed. They don't think it's necessary, because so much uranium is already available in the existing nuclear spent fuel, ripe for recycling.

"Frankly," Jake said, "there's enough energy content in the waste of today's reactors to power the whole country for 150 years."

The technology seemed so impressive—if it actually works and lives up to its promise—that I was a bit surprised when Caroline said that, for her, it wasn't the company's main selling point. "For us it's never been about the technology per se, we've never been technology zealots. We care a lot about the recycling," and fast reactors are more conducive to that, "but it's not like, 'Oh, one technology's so far superior.' It's really about the business model and how do you bring it to customers, and how do you interact with that community?" she said, sounding more like the purveyor of a neighborhood juice bar than Mr. Burns of *The Simpsons*. "What do you offer them?"

One possibility is that community customers could buy power directly from Oklo; the other is that Oklo could work with a utility. In some markets, utilities buy power from independent power producers such as wind and solar farms, and utilities could have the same kind of relationship with Oklo. "It's built very similarly to a renewables model," Caroline said, "except we can provide firm clean power"—that is, not intermittent. "We see a lot of interest from smaller cooperatives," Caroline added. "I think that's where renewables have been empowering and where we think fission can be empowering, too."

At the time of my visit, Oklo was arguably the furthest along of all the advanced nuclear start-ups, and they embodied both the potential and the challenges of the whole sector. They had customers lined up—a mix of small local utilities, government institutions like the Department of Defense, and data centers—and they had won several multimillion-dollar awards from the Department of Energy. And, crucially, Oklo was the first advanced nuclear company to submit its license application to the Nuclear Regulatory Commission. (A start-up called NuScale Power had submitted its application earlier, but its technology is more similar to the existing fleet—they are designing smaller versions of the light-water reactors.)

Caroline and Jake had, however, recently encountered a high-profile setback. In January 2022, after the herculean effort of completing their application to the NRC, it was rejected. The reason was, in the agency's words, "Oklo's failure to provide information on several key topics for the Aurora design." In a press release, an NRC staff member cited "significant information gaps in its description of Aurora's potential accidents as well as its classification of safety systems and components."[5] The denial was issued "without prejudice," meaning that Oklo was free to apply again.

They were caught off guard by the rejection. They had spent many hours discussing the finer points with the agency's staff—first in person, then on calls during the pandemic. If they were missing information, they thought the NRC could have done a better job of explaining the gaps in these conversations, rather than later denying the application outright.

To Jake and Caroline, their struggles also reflected broader issues. The NRC is accustomed to approving a certain type of reactor. (Even that is something of an overstatement since, because of the industry's struggles, the agency has approved only a handful of applications since it was established in 1975.) Now they are being asked to approve a range of different designs. "Everything's new," said Caroline. In theory, because of its inherent safety features, the Aurora should be safer than the current fleet, they believe. "You could almost leave it out in a field and it wouldn't hurt anyone," she said. They think this should make it easier to approve, but the novelty makes it harder. Jake saw this backwardness in the entire culture of nuclear energy. "It's really built around maintaining the status quo. And that obviously doesn't work when you're trying to do new things."

As much as they were clearly trying to reinvent nuclear for the 21st century, they also seemed deeply rooted in its past, both in terms of their effort to adapt the IFR and their grasp of early nuclear history. They were trying to resuscitate a vision that had foundered because of what Jake called "the misguided anti-nuclearism of the past." After we talked, I poked around the office. Outside the conference room, I saw evidence of the founders' nuclear nerdiness: a bookcase stacked with titles such as *Introductory Nuclear Physics*, *Nuclear Energy Synergetics*, and *Fundamentals of Fluid Mechanics*. Displayed on the wall near the restrooms, there were black-framed ads on yellowing paper from companies such as PG&E and

Union Carbide, in a vintage aesthetic: "Atomic electric power is here," one proclaimed. "A peacetime dream come true."

LONG BEFORE JAKE AND CAROLINE BEGAN THEIR WORK AT IDAHO NATIONAL Lab, a startling number of historic events took place there. It was not only home to the earliest development of both light-water reactors and fast reactors and later to the IFR (back when it was called Argonne-West); it was also the site of the only nuclear accident on U.S. soil with known fatalities. On January 3, 1961, at a small reactor the Army was testing, three workers died while attempting to restart the reactor after a shutdown. The accident prompted improvements in safety protocols at the lab and in the nascent nuclear industry. In short, while far out of public view—indeed, largely because it is far out of public view—this remote locale has been central to the American nuclear enterprise.

In May 2023, I arrived in Idaho Falls, a town of about 67,000. A river runs through the center of town; the water from the falls is used for hydroelectric power. When I took a walk alongside the river, I sat down on one of the benches placed at intervals on the path. Tulips and dandelions sprouted from the grass. As I watched the river flow by, I spotted robin redbreasts, swallows, and a man carrying a fishing rod.

The morning after I arrived, Sarah Neumann, the media relations manager at INL, met me at my hotel. We climbed into her government-issued blue minivan and drove to pick up Shelly Norman, an INL tour guide. Both women had grown up in Idaho, moved away, and returned to work at the lab, and both exuded an unusual degree of enthusiasm for their jobs; Norman said she would do hers "for free." As we rode out to the lab, they told me about their workplace. There are about 5,700 employees (they are contractors, though, not federal employees). "After the Cold War, a lot of the waste for building bombs came here," Neumann said. In addition to those contractors, hundreds more work on the waste cleanup.

"The purpose of the National Labs is to do things that industry won't and academia shouldn't," Norman explained as we drove. "So we do have a lot of work that maybe industry will pick up later. We develop it, we patent it, and maybe they'll take it and deploy it. . . . Academia can't because maybe they don't have the safety parameters."

"There's seventeen National Labs. We are the nuclear lab," said Neumann, with some pride. "There's some nuclear research that happens at other labs, but really the bulk of the fission research that's happening happens at INL."

They have the space to demonstrate full-scale reactors; they have the capability to fabricate and test fuel; and they have the facilities to examine the fuel after it's been irradiated. "Other labs can do part of it, but we're the only one doing all of it. So it's really attractive to industry to be able to come here," Norman added. Many of the current advanced nuclear start-ups have worked with the lab in some capacity—an example of how the government action of recent years is bearing fruit. Historically, there was not a well-established pipeline between the labs and the private sector, but the recent programs, especially Obama's GAIN initiative, have begun to change that.

Our first stop was the Systems Integration Laboratory, an industrial-looking space where I met Shannon Bragg-Sitton, director of integrated energy and storage systems. Her work involved researching how nuclear power could fit into the larger, decarbonizing energy system. As we've seen—as Dave Freeman and Ralph Cavanagh argued—integrating both nuclear and renewables into the grid can pose challenges. But, Bragg-Sitton explained, gesturing toward an infographic on the wall, these challenges are not insurmountable. Although it is difficult to quickly ramp nuclear output up and down to accommodate an influx of wind or solar energy, there are other possibilities. When a reactor's electricity is not needed for the grid, the reactor could pivot to desalination, or to producing hydrogen fuel. Nuclear produces heat, which is needed for many industrial processes, like steel and cement manufacturing, that are currently carbon-intensive. Integrating nuclear with renewables, Bragg-Sitton said, "we find that we have a really great tool set to more reliably and more resiliently meet grid demand at all times, but then have excess energy that we can use to decarbonize a lot of other processes, whether that be in industry or transportation." This struck me as one of the stronger arguments for having nuclear as part of the mix.

When we left the building, I asked about the trees blooming with the ambrosially fragrant white flowers. Sarah told me they were May trees. "Winter was, like, eight months long this year," she said. I had clearly come at the right time. And it was hard to avoid a cheesy metaphor as I talked with people throughout my tour: it was springtime in more ways than one at Idaho National Lab.

"It is such a great time to be in the lab, there's so much going on," said Norman.

"It really is," Neumann agreed. "In the nineties—administrations change, thoughts change—they kind of figured, we know everything we need to know about nuclear. We've got nuclear reactors going and yeah, you know, we don't need any more. We're done with that. And at that time they really were talking about closing up the lab, and just kind of stopping everything."

"The dark days."

"And now it's like, 'Oh, we still want to do stuff.'"

We resumed driving on a two-lane road, past field after field. I asked what kinds of crops they had around here. "Potatoes," Norman said. (The motto on the Idaho license plate is "Famous Potatoes.") Also barley—Anheuser-Busch has a malt plant in Idaho Falls. "This looks to me like alfalfa." She pointed out silos for irrigation, a potato cellar. In the sky ahead of us, I noticed a small, bright yellow aircraft swerving noisily: a crop duster. In front of us on the road was a truck packed with hay, pieces of which were continually flying out at our windshield.

When we crested a hill, we were able to see all 890 square miles of the government property. We began to approach our destination: the complex of labs and other facilities about 35 miles west of Idaho Falls.

"You're gonna see a lot of guns, a lot of guards, go through a lot of gates out here," Norman told me. Then I saw a sign on our right, as we neared the complex, that didn't exactly instill calm: "Use of lethal force authorized."

Each time we entered a new facility, I received a radiation protection briefing in advance: don't eat, don't drink, don't touch your face. Before leaving each building, I placed my hands into a machine that would detect any stray radioactive particles. As we walked around the complex, we could see the towering silver dome where the Experimental Breeder Reactor-2 had been housed and where the 1986 test of the Integral Fast Reactor had taken place.

The people I saw were overwhelmingly White men. (Among the younger ones, I noticed a trend I had also observed back in town: long, well-manicured beards.) This may have reflected the demographics of Idaho, but also of the nuclear field, a long-standing homogeneity that was proving slow to change. Some of the advanced nuclear start-ups were more diverse, and at INL, Neumann told me, diversity was a goal, but they clearly had a long way to go. (I would later learn more about their failures on this front.) At the cafeteria, though, I encountered evidence of the lab's efforts to adapt to contemporary sensibilities, in what was perhaps the most welcoming restroom I'd ever seen. On the wall next to the door was a sign with the merged symbols for male and female, as well as a wheelchair icon. In case you missed the point, the word "INCLUSIVE" appeared below these images. Inside, it was spacious, spotless, and well stocked with menstrual hygiene products.

OUTSIDE THE FUEL CONDITIONING FACILITY, WE MET ROBERT MIKLOS, one of the division directors at the Material Fuels Complex, a slim man with a white goatee. We entered the corridor adjacent to the operating area. In a small space, we stood before a glass display case with mock fuel assemblies inside. "We reprocessed a lot of fuel here over the years, and then we stopped because of the moratorium on recovering," he said—that is, President Carter's attempt to lead by example and cease reprocessing.

As of 2019, the facility has been reprocessing once again. Using spent fuel from EBR-2, they are producing a type of fuel called high-assay low-enriched uranium (HALEU) that a number of the new start-ups are planning to use. HALEU is more highly enriched than the fuel used in the current fleet, which consists of about 3 percent U-235, but far less enriched than weapons-grade uranium, which is more than 90 percent U-235; HALEU is enriched to between 5 and 20 percent. There are several advantages of the higher concentration, including that fuel assemblies, and therefore reactors, can be smaller, and refueling can be less frequent. As always, though, there are also trade-offs: some nuclear safety experts have serious concerns about the risks of reprocessing and of the higher levels of enrichment.

Gesturing at a photo in the display case, Miklos said, "That's recovered uranium right there. It kind of looks like tree moss. That's pure uranium." I peered at the photo. To me it looked like a clump of black shredded paper. Once they have the pure uranium, the next step is to "downblend" it by adding depleted uranium. Then it is turned into a solid metallic form. "This is actually gonna be the feedstock for the new fuel we're going to fabricate for Oklo," he told me.

Now it was time to enter the operating area. They had recently had a planned power outage, so they were not currently operating. If they had been, Miklos told me, I would have seen technicians all around. Inside was a "hot cell," 16 feet across and 26 feet high—a protected room where radioactive materials are contained. The walls are five feet thick, as are the windows, which are leaded to prevent the escape of radiation and oiled to make the objects inside look closer than they are. The room is also filled with argon rather than air, for an inert atmosphere that won't corrode the materials.

Inside the hot cell, it was murky and dim, and robotic arms extended from the walls. Operators undergo training for up to two years to get qualified to use these, to manipulate the radioactive materials in the hot cell remotely. "Some people become very proficient, others are not," said Miklos. "It's amazing what the operators are able to do—use screwdrivers, little Allen wrenches. I know I couldn't do it."

Other than the periodic outages, they have been running 7 days a week, 24 hours a day—with two shifts, operators working 12-hour days. This is because of the extraordinary surge in demand. "There's a real resurrection of energy initiatives here in the United States," he went on, thanks to all of the nuclear start-ups that had sprouted up in the past decade or two, and the support for them from the GAIN initiative and Congress. "That has made the demand for this. You know, when there weren't start-up companies and new advanced reactor programs, and a decrease even in commercial reactors, the demand side wasn't there." Another driver was the urgency of ending our dependence on Russia for fuel. "What we've done is we've eliminated what used to be just a liability sitting in storage, and now put it back in the fuel cycle," Miklos said. "For the folks working here and myself, it's just fantastic. You took what was considered to be a waste and turned it into a national asset."

At Oklo's offices, and in Idaho Falls, it was easy to get caught up in the enthusiasm. Everyone I met was clearly excited about their work, excited about the future of nuclear. They were motivated by the joy of the scientific innovation and by the prospect of contributing to the good of the nation and what used to be called "mankind."

But we have to step back and look soberly at the challenges faced by advanced nuclear. The reactors have repeatedly taken longer than expected to become a reality, and they are still unproven. To renewables advocates like the late Dave Freeman and Stanford professor Mark Z. Jacobson, it is sheer folly to invest in them when we could be investing the same money and effort into building out wind and solar and improving storage and efficiency. And it bears emphasizing how controversial reprocessing is. Recycling seems like a no-brainer in theory, but several analysts I spoke with were adamant that it increases the likelihood of weapons diversion and is not worth the risk or expense.[6]

At the lab, aside from the vibe of exuberance, what struck me most were the technical complexity of the work occurring there and the security—the guards, the guns, the sign about "lethal force." As at the other nuclear sites I visited, the security measures were a stark reminder of the unique risks associated with nuclear and the kind of heavy-handed state power that inevitably goes along with it. "The problems of nuclear power, in contrast to other energy technologies, are institutionally serious as well as technically unresolved," wrote Ralph Nader and his coauthor, John Abbots, in a 1977 book, *The Menace of Atomic Power*. "Nuclear power tempts saboteurs, terrorists, and hostile nations."[7] At the same time, it occurred to me that part of the reason some people are so drawn to nuclear is precisely because it's hard, because it's powerful, maybe even a little bit because it's dangerous. Having unlocked the secrets of nuclear physics, or at least begun to, how can we not pursue them? It's obviously a very human trait to want to know, to explore, to invent, for better and for worse. Speaking for myself, these aspects partly explain my own fascination. I think solar is great, for example, but I can't imagine that it would sustain my interest enough to write a book about it.

My fascination comes with a healthy dose of ambivalence. On some level, I still pine for the simple, wholesome, lower-tech world that has long been an environmentalist dream. But it must also be said that both the complexity and the security may trouble me less than they did my antecedents in the anti-nuclear movement. That's because nearly everything is complex now, and security, albeit not quite to the same degree, is everywhere; there are armed guards at my daughter's school. Just as the waste doesn't seem as disturbing because I'm aware of so many other public health risks, nuclear in general no longer evokes singular existential dread for me. This is at least as much a lamentation of the modern world as a defense of nuclear, but I suspect it explains something psychological about the recent shift in attitudes.

After my visit, the advanced nuclear sector continued to experience ups and downs. Oklo continued to attract interest from potential customers, and continued to work with the NRC toward resubmitting its license application. But a few months later, in November, came a warning for the entire field. NuScale, the company working on small modular light-water reactors, had made a deal to supply power to the Utah Associated Municipal Power Systems, a collective of 50 utilities in seven states. But when NuScale significantly revised its cost estimate upward, that deal fell through. The failure raised new questions about the financial viability of advanced nuclear. Hanging over all of the setbacks was the question: how much of the difficulty was due to the intrinsic nature of nuclear energy, and how much was due to regulation that was in need of reform? Regardless, it was clear that a future in which small modular reactors dotted the landscape was coming along more slowly than their champions had hoped.

CHAPTER EIGHTEEN

THE TURNING POINT

IF THE FLEDGLING PRO-NUCLEAR MOVEMENT HAD TRIED TO INVENT A NEW avatar for the cause, they could hardly have come up with someone better suited to the task than Isabelle Boemeke. In 2020, she arrived on nuclear Twitter like a godsend. While other advocates each contributed something new and valuable to the cause—making it fun, showing that not all nuclear advocates were White men, communicating with progressives—Isabelle combined all of those assets. And she brought something no one else had: glamour.

Isabelle had grown up in the '90s in small-town Southern Brazil, where people sometimes still rode horses in the street. She lived in a series of small apartments with her mother. In the summer, the heat was intense. Since they didn't have air-conditioning, they had to open the windows, but then mosquitoes swarmed inside. In the winter, the temperatures would dip into the 40s. They owned one electric heater, and Isabelle would limit her use of it, conscious of electricity costs. She would wear multiple layers of clothes to sleep, and she often skipped showering because it was so cold. Her grandparents would fill metal buckets with alcohol that they would light on fire, then put the flaming buckets in the bathroom to warm it up. "Like, this is a thing that people did," Isabelle marveled when we spoke.

By high school, she and her mother had moved to the city of Curitiba (known as the "green capital" of Brazil for its pedestrian-friendly streets and public transit). Isabelle had dreamed of being a psychologist. But she also happened to be very pretty. One day, in the winter of 2006, when she was 16 years old, a man approached her as she was leaving school. He asked if she was a model. "And I'm like, 'No.' And he's like, 'Do you want to be one?' And I'm like, 'Maybe.'" She saw it as her ticket out of Brazil.

There was a contest happening the next day in Curitiba. Isabelle ended up placing third in the national competition and secured a contract with the agency. She began modeling, first in São Paulo, then Argentina. Eventually, in 2008, she made Miami her home base and, from there, traveled all over the U.S. and Europe.

When Isabelle was 19, she read *The Greatest Show on Earth*, a book about evolution by Richard Dawkins, and grew interested in science. On Twitter, she followed a number of scientists, including Carolyn Porco, a planetary scientist who had worked closely with Carl Sagan. Sometime in 2015 or 2016, Isabelle noticed a tweet from Porco about molten salt thorium reactors—a kind of advanced reactor that would use thorium as a fuel and molten salt as a coolant. This was one of the reactor designs that people got excited about and thought would solve the problems associated with current reactors. It piqued Isabelle's interest, especially given her early experiences of inadequate energy access. "It was like, 'Oh, this sounds interesting. I've never heard anybody say anything positive about nuclear. I don't even know anything about nuclear. This is cool.'"

Up to that point, Isabelle had been somewhat concerned about climate change, but she thought, *The adults in the room are taking care of this.* Her wake-up call came in 2019, when bushfires ripped through Australia, eventually burning about 243,000 square kilometers and killing more than a billion animals. The same year, wildfires in the Amazon, in her home country, destroyed an area the size of New Jersey. By 2020, she was living in the Bay Area, and wildfires in California had been turning the skies orange. "That was my moment of, 'Okay, we're not moving nearly as fast as we have to. The adults are definitely not in the room . . .' and I just felt this desire to do whatever I could to help address this problem."

She began looking more seriously into nuclear, not just molten salt thorium reactors but the kind that already existed, asking scientists about this form of energy. As she would later report in a TED Talk, they consistently told her: "It's good. We need it. People hate it." She asked herself, "'What can I, with my very limited skills as a fashion model, do to help some small part of this gigantic puzzle?' It was more, like, out of despair really. I can't just live my life as if nothing is happening." One day when she was brushing her teeth, she told me, an idea came to her: nuclear influencer. Instead of just posting selfies and videos of herself applying makeup, she could post videos about nuclear energy. She would conduct a fun bait and switch: the videos would start out as beauty tutorials, then pivot to science communication.

Around this time, in her quest to learn about energy, she met with a well-known expert named Saul Griffith. He was an Australian inventor and entrepreneur—he'd founded a company that developed airborne wind turbines—and had won a MacArthur "genius" award in 2007. When she met with him at his office in San Francisco, she floated her idea: she was thinking of becoming a nuclear influencer. Griffith was not a big fan of nuclear, though not vehemently against it. He thought it would be helpful to have some in certain areas, but that renewables could meet most of the demand—and in his 2021 book *Electrify!*, he would lay out in detail how to make that happen. Still, after Isabelle left his office, he sent her an email with a suggestion. If she did go ahead and try to become a nuclear influencer, she should call herself "Isodope."

Thus Isodope was born. One of her first TikTok videos begins, "Hey, guys, so a lot of you have been asking about my skin care? I love this mask, not just because it looks like I'm sleeping in a pod on a rocket on my way to Mars. But also because it makes my skin look so radiant. Radiant like uranium pellets." Then she goes on to say that radiation is everywhere and that we shouldn't be afraid of it. The video is full of saturated colors and quick cuts. In every frame, you see Isabelle's light-blue eyes, thick dark hair parted down the middle, and, yes, radiant skin. She went on to make videos about various other aspects of nuclear energy. One is an "ASMR" video in which she whispers huskily and makes sounds by playing with pink putty. "Hey, guys, this is my first ASMR video," she says. "'ASMR' stands, of course, for 'a small modular reactor.'"

The existing pro-nuclear community—Heather and Kristin, Eric Meyer, Paris Ortiz-Wines, the Breakthrough Institute, and others—was thrilled by these videos, which also found an audience beyond that core group. Isodope's platform at the time was relatively small—she had tens of thousands of followers, not hundreds of thousands or millions—but she was reaching people who had never thought about nuclear energy before. Isabelle also started to ponder Diablo Canyon, the plant not far from where she now lived. She made a video about it, wearing a shiny black suit with the word "DISCO" emblazoned on it over and over again, her bangs in two braids. "When a nuclear plant is shut down, it is replaced by fossil fuels," she said.

As well as glamorous and beautiful, Isabelle was, not coincidentally, extremely well-connected. Her romantic partner was Joe Gebbia, a founder of Airbnb. She was friends with Elon Musk and close to the musical artist Grimes, Musk's girlfriend at the time.[1] In December 2021, Grimes and Isabelle made a video together. Grimes, wearing what appears to be a blond wig and purple and yellow face paint, speaks to the camera while Isabelle sits to her right, wearing dark sunglasses and reading a book. "Hey, everybody, it's Grimes, and I'm here to talk about Diablo Canyon and nuclear power in California," she said. "Closing Diablo Canyon will make us reliant on fossil fuels. This will push the state backward instead of forward in its goal to become one hundred percent reliant on clean energy. . . . This is crisis mode, and we should be using all the tools that we have, especially the ones sitting right here in front of us." Then she blows the audience a kiss.

Isabelle, then, was pretty much the opposite of Gene Nelson. He labored largely behind the scenes, submitting his hundreds of pages of comments to the California Public Utilities Commission, with a minimal internet presence and little connection to the zeitgeist; nobody would accuse him of being stylish.[2] She, by contrast, was working in a very public way to make nuclear, and Diablo Canyon, as fashionable as she was. In these ways, the two advocates were complementary. But what made Isabelle even more valuable to the movement was that she did not limit her activity to snappy videos; soon after she started her social media activities, she decided she also wanted to do more on-the-ground activism. Before long, she connected with Paris Ortiz-Wines through Twitter, and Paris brought Heather Hoff into their conversations.

Mothers for Nuclear had been plugging along, giving presentations at nuclear conferences and local schools, posting on Twitter and Instagram, going on *Dave Congalton Hometown Radio*. They felt like they were having good conversations about nuclear energy, but when it came to Diablo Canyon specifically, they were fairly pessimistic. Every passing week—as decommissioning planning continued, as equipment wasn't replaced because the plant was going to close anyway—it seemed less likely that a reversal would occur. Isabelle's energy and celebrity, however, helped change the dynamic. It was the depths of the pandemic, but they began holding weekly Zoom meetings to revive the campaign, to plot how to save Diablo Canyon.

Isabelle told me she knew very little about the first wave of activism to save the plant, back in 2016, when the deal to close it was announced. "I knew of Shellenberger, obviously," she told me, "and I knew that he had done some effort on Diablo Canyon, but I knew that it had failed, and that was all I knew."

Yet, knowingly or not, she largely replicated what Shellenberger had done initially. She founded a nonprofit, called Save Clean Energy, which was basically just her and a consultant she hired. She decided to try to get scientists to sign an open letter to the governor urging him to reverse course, just as Shellenberger had done six years earlier. It was released on February 1, 2022, with 79 signatories, including a number of familiar names: Steven Chu, Kerry Emanuel, James Hansen, Carolyn Porco, and David Victor, as well as other professors at Stanford, Yale, Princeton, and MIT. "Considering our climate crisis, closing the plant is not only irresponsible, the consequences could be catastrophic," the letter read. "We are in a rush to decarbonize and hopefully save our planet from the worsening effects of climate change. We categorically believe that shutting down Diablo Canyon in 2025 is at odds with this goal. It will increase greenhouse gas emissions, air pollution, and make reaching the goal of 100 percent clean electricity by 2045 much harder and more expensive."[3]

Meanwhile, Isabelle was working with her new collaborators, who happened to be Shellenberger's former collaborators, to organize a rally. On December 4, 2021, they gathered in downtown San Luis Obispo, on the green in front of the courthouse there. It was reminiscent of the march in San Francisco in 2016: they had some of the same chants ("Save the plant! Save the plan-et!") and similar signs. Gene Nelson was there, in his green headband; so was Scott Lathrop of the YTT Northern Chumash Tribe. Isabelle led the march, wearing a Save Clean Energy T-shirt. This time, all of the speakers, including Isabelle, Heather, Kristin, and Carolyn Porco, were women—a deliberate choice to highlight female advocates. One twist was that Isabelle rented a blimp for the occasion. On the side it read, "This blimp represents 1 tonne of CO_2. Replacing Diablo with natural gas would add 7 million tonnes of CO_2 to the atmosphere." "Listen to the scientists!" the crowd chanted.

Just as Shellenberger was stepping back from the movement, Isabelle stepped in. Some newer pro-nuclear advocates had no connection to him and wanted nothing to do with him. Isabelle shared some of Shellenberger's assets: she, too, was telegenic and a gifted communicator. But she was more diplomatic, with more of a sense of humor. As Shellenberger cemented his position as a darling of the right, Isabelle, who considered herself progressive at the time, courted a different audience.

In April 2022, she gave a TED Talk recounting her conversion. And she made explicit what she had been trying to do—what she had been succeeding in doing. "I wanted to make nuclear energy cool," she said in her gently accented English. "Now, don't get me wrong, I love renewables, but to me, it's clear that we need more. We need a source of energy that's clean and works twenty-four-seven to complement them." She drew on the concept of the "meme," first introduced by Richard Dawkins, referring simply to an idea that spreads through the culture. Anti-nuclear bias, Isabelle proposed, is nothing more than a bad meme. After reviewing the arguments against nuclear—and shooting them down—she ends with this: "I'm not asking you to build a reactor. I'm not asking you to rewrite *The Simpsons*. I'm not even asking you to sign a petition on change.org. All I'm asking you is to please join me in spreading the meme: Nuclear energy is cool."[4]

The audience members cheered, and some stood. When I last checked, the talk had more than 1.7 million views.

While Third Way had been making remarkable inroads on Capitol Hill, the Clean Air Task Force in Boston had also been working quietly in the background, producing some of the research and analysis that Third Way could point to in its lobbying efforts. Since its founding in 1996, the executive director of CATF had been Armond Cohen, a mild-mannered Harvard Law School alum. He had strong connections with others in the Boston area who also supported nuclear and other technological solutions to climate change. In the summer of 2020, he heard from one of them, Jacopo Buongiorno, a professor of nuclear engineering at MIT.

Diablo Canyon had become a cause célèbre for the pro-nuclear community throughout the country, even the world, because of its symbolic value. With its location on the California coast, where whales could sometimes be seen breaching not far from the shore, it had become the plant whose image was nearly always used in favorable articles about nuclear; Alex Trembath of the Breakthrough Institute described it aptly to me as "charismatic."

Buongiorno was playing around with ideas to demonstrate the value of Diablo Canyon. Cohen put Buongiorno in touch with Sally Benson, a professor of energy science and engineering at Stanford. In November 2021, they, along with six coauthors, released a 114-page report titled "An Assessment of the Diablo Canyon Nuclear Plant for Zero-Carbon Electricity, Desalination, and Hydrogen Production."[5] This was not a peer-reviewed study but a report funded by two centers at MIT, including the Center for Advanced Nuclear Energy, and several of the major pro-nuclear donors, including the Pritzker Innovation Fund, the Rothrock Family Fund, and

Ross Koningstein. (As for Buongiorno, he had a title that may raise eyebrows: his professorship at the time was an endowed chair sponsored by TEPCO, the Tokyo Electric Power Company, which owned the plant in Fukushima. He told me that there were no strings attached; the endowment was established in 2000.)

The report projected that keeping Diablo Canyon until 2035 (10 years later than it was currently scheduled to close) would reduce carbon emissions from the state's power sector by more than 10 percent from 2017 levels; save $2.6 billion; and increase the reliability of the grid. If the plant's life was extended to at least 2045, the projected cost savings would rise to $21 billion, while an estimated 90,000 acres of land, which would otherwise be needed for renewables, would be spared. In addition, deploying the plant for desalination—removing salt from seawater—could "substantially augment freshwater supplies to the state," and a hypothetical hydrogen plant connected to Diablo Canyon could produce zero-carbon hydrogen fuel, which could be used to power vehicles. The authors envisioned the plant as a "polygeneration facility" that would provide these services in a coordinated way, offering massive value to the state.

They also hired a PR firm and set up a series of meetings with different stakeholders: Governor Newsom's office, state representatives, unions, newspaper editorial boards. In November 2021, both of Obama's former secretaries of energy, Steven Chu and Ernie Moniz, wrote an op-ed in the *Los Angeles Times* that cited the report and concluded, "Reimagining Diablo Canyon's role in California's energy future is an opportunity we cannot afford to ignore."[6]

Buoyed by this response, Cohen decided to try to foster a California-based coalition to start advocating for the plant. He helped launch a 501(c)(4) called Carbon-Free California. It was a loose coalition of tech entrepreneurs and labor people that Cohen brought together, primarily with tech funding. (They did not take any funding from energy interests.) "We found some entrepreneurs in the Valley that wrote some very large checks," Ray Rothrock, a board member, told me. They quickly raised more than $3 million.

Some of those in the coalition had connections in the Newsom administration. When they reached out to their contacts, "at first they were sort of standoffish," said Cohen. They seemed to think, "'we don't need this headache,' and PG&E didn't want to go there at all." Early in 2022, Carbon-Free California hired Newsom's pollster and showed his staff what they found: that 58 percent of California voters supported extending the life of the plant; in San Luis Obispo County, 74 percent of voters did. The staffers were "quite sympathetic on the merits," Cohen said, but not ready to make any commitments.

In 2019, the YTT Northern Chumash Tribe had started picking up on signs that plans for Diablo Canyon could change. First, Scott Lathrop says, they received calls from three different entities that expressed interest in purchasing the plant from PG&E. (While Gene Nelson and his group had been looking for a buyer, it is not clear whether they were involved. Scott declined to name the interested parties.) According to Scott, these entities approached the tribe and said, "Hey, confidentially, we're doing our due diligence, and we're wondering what your opinion is about the plant going forward because we're thinking about buying the plant."

These offers ultimately fizzled, but then Scott got another call: from Isabelle Boemeke. He met with Isabelle, Heather Hoff, and another young advocate, Ryan Pickering, at a café in downtown San Luis Obispo. The activists wanted to hear Scott's point of view and be supportive of the YTT; they were also hoping the tribe might make a statement in support of continued operations. But the tribe was not willing to make that statement—they wanted to keep their focus squarely on recovering the land—and Scott conveyed that clearly.

In their conversation, though, he was struck by how well-connected Isabelle seemed to be. When she asked him, "What can we do for you?" he said, "What you can do is connect us with any and all players. Whether that's in the nuclear industry, government." She did, later facilitating meetings with the president of the American Nuclear Society (ANS) and a prominent Washington lobbyist, Kai Anderson. Scott wanted to make sure that no matter what happened with the plant, his tribe could have a seat at the table and their interests would be considered. In his meeting with the ANS president, he said, "If you don't want a bunch of stink from the Natives, you need to understand this and push for including this landback because it's the right thing to do."

At the same time, Scott was still on the Decommissioning Engagement Panel, and he visited the plant a number of times in that capacity. It looked to him like it was in excellent condition. He began to think, "This doesn't seem to make a lot of sense, to tear this down."

Based on his observations—the growing interest from various quarters in preserving the plant, the fact that it seemed to be in good shape—he was increasingly skeptical that it would shut down as planned. He told his tribal group, "Hey, folks, mark my words, that plant is gonna go forward. We can either be in the mix or be outside the mix, but I believe it's gonna go forward."

DIABLO CANYON WAS THE SINGLE LARGEST SOURCE NOT ONLY OF LOW-carbon electricity in the state but of any kind of electricity, supplying about 9 percent of in-state generation.[7] In 2020, this contribution took on new salience. On August 14 and 15, for the first time since the energy crisis in 2001, the state grid operator, known as CAISO, instituted rolling blackouts, largely due to a brutal heat wave, which led to greater use of air-conditioning. The situation highlighted how energy use and climate change feed on each other in a vicious cycle. The blackouts affected about 800,000 homes and businesses, though no one lost power for longer than 2.5 hours.[8] A few weeks later, the grid operator came close to having to impose rotating outages again, but ultimately managed to keep the lights on this time.

For Newsom, this appears to have been a wake-up call. He was acutely aware that Gray Davis, the Democratic governor who had presided over the blackouts two decades earlier, had been recalled in a special election in 2003. Newsom was widely known to be harboring national political ambition, and he could not afford to have blackouts again on his watch.

In the next couple of years, the effects of climate change continued to strain electricity supplies. A drought diminished hydro resources, and wildfires interfered with transmission lines. At the same time, California was attempting to accelerate the transition to clean energy, which would mean vastly increasing electricity use. Just a month after the 2020 rolling blackouts, Newsom issued an executive order mandating that by 2035, all new cars and passenger trucks sold in the state be zero-emission—that is, electric. And yet, the state's plan was to shut down its largest single source of electricity.

On April 29, 2022, the *Los Angeles Times* broke the news that the previous day, Newsom had told the newspaper's editorial board that he was reconsidering the decision to close Diablo Canyon. Earlier that month, the Biden administration had announced the $6 billion pot of money to bail out economically struggling plants. Newsom said, "We would be remiss not to put that on the table as an option."[9]

Newsom told the *Times* that he had been thinking of extending the life of Diablo Canyon since the rolling blackouts of 2020. He also mentioned the advocacy from scientists and activists, apparently alluding to Isabelle's letter and the op-ed by Moniz and Chu. "Some would say it's the righteous and right climate decision," he said.

It was a stunning turnabout. The pro-nuclear camp gave Newsom lots of love on social media, and any opposition was fairly muted. As Armond Cohen of Clean Air Task Force saw it, "He floated a trial balloon . . . and the balloon didn't get shot down completely out of the sky."

A MONTH AND A HALF LATER, I DROVE TO THE ANNUAL MEETING OF THE American Nuclear Society, at a Hilton hotel in Anaheim, the Southern California city where Disneyland is located. The ANS is an organization representing engineers, scientists, and industry professionals in the nuclear field. It had historically been rather staid and apolitical, but had recently edged more into advocacy. At lunchtime on the first day of the conference, there was a "show of support" for Diablo Canyon outside the hotel, amid the concrete and palm trees. A few dozen attendees, dressed in business casual attire, with their name-tag lanyards around their necks, held two banners with the ANS logo, and a graphic of a light bulb superimposed on an image of the plant. One banner read "Keep Diablo Canyon running!" and another read "Nuclear energy = clean energy." Standing in front, panning the scene with his phone, was Gene Nelson, who was wearing a dark suit, black backpack, and his green headband. Gene would also be receiving an award in recognition of his thousands of hours of work on behalf of the plant.

As at the last nuclear conference I had attended, a few years earlier in San Diego, most of the panels were highly technical, on subjects like reactor physics and thermal hydraulics. But one, starting at 10:15 on the second morning, was titled "Diablo Canyon and California: Facing the realities of California's energy challenges?" The panelists included Heather Hoff, Gene Nelson, Scott Lathrop, and—in what was probably the first appearance by a fashion model at an ANS meeting—Isabelle Boemeke. Joining virtually were Jacopo Buongiorno (coauthor of what had become known as the Stanford-MIT report), a San Luis Obispo County supervisor, and a leader of the International Brotherhood of Electrical Workers, the union that represented most plant workers.

Before the panel started, we milled around drinking coffee in a generic conference room. Heather was dressed up in a skirt suit. She introduced me to Isabelle, who was friendly and down-to-earth, wearing baggy pants and chunky boots. On her phone, Heather showed me that just that morning, Senator Dianne Feinstein had published an op-ed in the *Sacramento Bee*, headlined "Why I changed my mind about California's Diablo Canyon nuclear plant."

"If California is to lead the clean energy transition, as state law mandates, Diablo must keep operating, at least for the time being," she wrote.[10] At this point, Feinstein's mental fitness was already in question, and she was likely outsourcing a lot of work to her aides. It didn't matter. The point was that the Democratic establishment was closing ranks, forming a consensus that Diablo should live on.

The panelists sat in cream-colored chairs on the stage, in front of a velvety blue curtain: Scott on the left, then Heather, Gene, and Isabelle. To Isabelle's right, a large screen showed the three panelists who were joining virtually.

When Scott delivered remarks, from the lectern on the left of the stage, he was almost comically diplomatic and understated. "In the early seventeen hundreds, we had kind of an interesting experience with the Mission system," he said. "There's been a lot of interesting things that have happened to our homeland, that we have not had the opportunity to be the direct decision-makers or have that input." He mentioned that Diablo Canyon had been built on top of his ancestors' burial sites, which was "not a real good thing for the tribal community." But, he went on, "things happen . . . and now we ask the question: Is Diablo Canyon a liability or is it an asset? And from our tribal position, we would take a look at Diablo Canyon as more of an asset." They wanted to have some say in the plans, though, and they wanted to recover their land. "We believe as a tribal group that we need to be in the middle of the decision-making process."

When it was Heather's turn, she said, "We need more clean energy. It's obvious perhaps more now than ever before, with the situation in Ukraine and here in California. We don't have enough electricity, period." Gene gave a history lesson about reliability ("Let's review what happened during the Enron crisis in 2000–2001. . . . So now we're going to look at a near-miss CAISO stage 2 emergency caused by the Bootleg fire on the evening of Friday, July 9, 2021.")

Isabelle went last. She started with a land acknowledgment of the YTT Tribe. Then she said, "Diablo Canyon is so much more than California's last nuclear power plant. I would actually say that it is the most politically relevant plant in the United States. California's obviously known as a leader in the fight against climate change, and Diablo Canyon, as the state's largest source of not only clean electricity but electricity in general, should be at the center of any serious decarbonization strategy."

Most of the panelists seemed almost giddy that it seemed like their long-shot goal might be within reach. As Isabelle put it, "What seemed impossible now seems feasible, and maybe even likely."

CHAPTER NINETEEN

A SOCIAL AND POLITICAL PROBLEM

In the summer of 2018, Charlyne Smith drove with her father from Gainesville, Florida, to Idaho Falls. She had just completed her first year of a doctoral program in nuclear engineering at the University of Florida, and she was about to start an internship at Idaho National Lab. Her father was a professional truck driver, and he wanted to do the 35-hour drive in one go. They stopped only for gas and, once per day, at a rest stop to nap for a few hours in the car.

In her doctoral program, Charlyne was working on a research project called "Fission Gas Pore Size Distribution in Uranium-Molybdenum Fuels." It involved studying a kind of low-enriched fuel that could reduce the proliferation risk of research reactors. For the project, she had been awarded a prestigious National Science Foundation fellowship. At the lab's world-class facilities, she would be able to examine the fuel after it was irradiated.

Charlyne was from Jamaica, and she was full of ideas about how nuclear energy could be deployed to benefit her country and the rest of the Caribbean. With a friend, she had founded a start-up focused on developing floating small modular reactors to send to island nations whose power was out as a result of natural disasters. She had recently met Eric Meyer, of Generation Atomic, when presenting a poster at a conference at her university, and she liked that he was trying to talk to people about nuclear in terms of personal stories. "I'm in nuclear because I realize the potential that you can have on people's lives, ultimately," she told me when we first talked. She would later volunteer to become a "nuclear ambassador" for Generation Atomic, attending climate and energy conferences and talking to random people about nuclear.

In the late 2010s, the American nuclear industry and culture still seemed, in many ways, like a relic. There were lots of ideas about how the field needed to adapt to the 21st century. A new camp was emerging in the pro-nuclear space, to focus on the imperative for change: it needed to become more diverse and welcoming; it needed to engage more respectfully with communities; it needed to incorporate social science insights into its operations and public outreach. Others were focusing on regulatory reforms that could make advanced nuclear easier to build.

Charlyne herself embodied some of these incipient changes. Part of it was by virtue of who she was: she was on her way to becoming the first Black woman to earn a doctorate from the nuclear engineering program at her university. She would also end up pursuing work focused in part on changing the regulatory landscape. But it wouldn't always be easy. The entrenched nature of the status quo, the very reasons she represented change, also sometimes made her journey a bumpy one.

But as she and her father approached Idaho, the ride was smooth. The road trip was the first time she'd gone from state to state, seeing the welcome signs on the highway. Her father was driving when they arrived in Idaho, and she was asleep, but she'd asked him to wake her up when they crossed the border from Wyoming. "It was just the most gorgeous scene I'd ever seen," she said. "It's not like Jamaica gorgeous. It's a different type of gorgeous there. You're driving, and it feels like you're driving into a painting." She could see the mountains, with residual ice on top, even in June. "And everything just looks so . . . otherworldly."

Charlyne grew up in a small town called Old Harbour, about 25 miles west of Kingston, near Jamaica's southern coast. Her mother worked for the Jamaican equivalent of child protective services, and her father was a jack-of-all-trades. (One of his stints was managing a fish farm; sometimes he would catch crocodiles that were preying on the fish and take them to the zoo.) But, in search of more stable employment, he left for the U.S. when Charlyne was about five. Working construction jobs under the table there, he sent money home to his family: Charlyne, her mother, and her older sister and two younger brothers.

Like Isabelle Boemeke, Charlyne grew up without many modern amenities. In her house, the faucets of the sinks and shower each had a red knob that was supposed to represent "hot," but nothing would come out; apparently it just was there for show. Early every Saturday morning, her mother would wake up all the kids at dawn and hustle them outside, bleary-eyed. Each of them would get a plastic bin to wash their own undergarments and school uniforms. Their mom would fill the bins with water and tell them to start scrubbing. They used a grainy cleanser called cake soap; she told them they weren't washing properly if they didn't hear the *scroop-scroop* sound. After they washed each item in the bin, they'd wring it out and rinse it in another bin. After rinsing twice, they got a clothespin and hung the garments on the clothesline. Then it was time to clean the house.

The family did have electric lights and a refrigerator, but the service was iffy. The lights could go out at any time for short periods, and when hurricane season came, in June of each year, the outages were sometimes prolonged. To prepare, the family would put up tarpaulin around the house, board the windows, get their kerosene oils ready, and stock up on canned food. Sometimes the power would be out for days. After dark, they would sit around the dinner table, which was in the center of the house, away from the windows. The only light came from kerosene lamps and candles. The wind sometimes sounded like voices, which would frighten Charlyne and her brothers, and her older sister would tease them. They didn't know when the power would return; there was nothing to do but wait.

When Charlyne was a teenager, she came to the U.S. for the first time, to visit her father in Pennsylvania. The trip was a revelation. She had never liked showering because the water was always so cold. In Pennsylvania, she discovered the pleasure of hot showers. "I was in the shower *all day*," she told me. "The bathroom was steamy. Outside the bathroom was steamy. I was wasting all this water because I'm like, this is amazing." There was a washing machine. There was a dryer. There was a dishwasher. There was even a pancake maker. "I mean, I was in paradise."

Years later, she would hear the arguments for the importance of energy access: it's not just that it bestows conveniences and luxuries but that it can open up space for deeper values, contributing to, for example, higher literacy rates and gender equality. In 2010, Hans Rosling, the late Swedish physician and author, elucidated these benefits in a TED Talk titled "The Magic Washing Machine." He stood onstage next to a washing machine, explaining how it had changed his mother's life and his own. When she no longer had to spend hours washing clothes by hand, she would load the machine and take him to the library. "This is the magic," he said. "You load the laundry, and what do you get out of the machine?" He reached into the machine and pulled out several hardcovers. "You get books."[1]

On that first trip to the U.S., Charlyne likewise marveled at the amount of time saved when she didn't have to wash her clothes or dishes by hand—although she admits she didn't necessarily spend it in any grand, ambitious way. After all, she was a teenager. "I'm gonna watch TV, you know?" She started to wonder, was her father rich or did everyone in the U.S. live this way? "I begin to realize that, everybody got it! It's not something that's for the elite, because that's how it is in Jamaica."

Charlyne had been attending a performing arts school; she was in the choir, the drama club. But after her visit, as she started to think seriously about her future, she began to gravitate toward science and technology. Part of it was sheer self-interest: she wanted to achieve financial security so her family could enjoy the amenities she had learned were widespread in the U.S. But at some point, she started thinking bigger: she wanted to help her whole country access these benefits.

By the time she graduated from high school, in 2012, her father had become a U.S. citizen, which allowed her to join him. In the fall of 2013, she matriculated at Coppin State University, a historically Black college in Baltimore, where she majored in chemistry and mathematics and served as president of the STEM club. At first, she was interested in studying solar energy. Climate change was not on her radar at all, but solar just seemed to make sense for Jamaica, because it's an island with lavish sunshine. (Most of Jamaica's energy came from imported fossil fuels.) She began participating in a solar project, which involved detoxifying the process for manufacturing panels by using fruit dyes instead of chemicals. The research was interesting, but Charlyne was looking for something that would have a transformative impact, and she began to have doubts about solar, especially after learning about its intermittency.

Around this time, the STEM club hosted a speaker named Nickie Peters, a nuclear engineer who worked at the University of Missouri research reactor. He also happened to be from the Caribbean. Charlyne really knew nothing about nuclear at that point and had no preconceived notions about it. Intrigued, she went to talk to Peters after his presentation. She was especially struck by his comment that it could be used not just for electricity generation but for other energy-intensive applications, such as desalination, which she thought might be helpful in addressing Jamaica's water issues. She later learned that Jamaica had a small research reactor—the only country in the region that had any type of reactor. After doing more research on her own, she began to imagine a Caribbean powered by nuclear energy, she told me. "That was my vision from then."

WHEN CHARLYNE ARRIVED IN IDAHO FALLS, SHE MOVED INTO THE BASEment of two Italian staff scientists her advisor had connected her with. Her third day of work, these Italians drove her to "the site," the complex about 35 miles west of Idaho Falls where most of the facilities are located. On the way, she noticed the sign that would later catch my attention: "Use of lethal force authorized." She thought, *I'm gonna be doing some serious things here. I don't know if I should be excited or scared.*

After arriving and going through a security checkpoint, she was given a dosimeter, to record her radiation dose wherever she went. It was going smoothly, she was taking it all in—the scenery, the buildings, the people. "It just felt like it was gonna be different," she said. "That on top of the fact that"—she paused for a few moments—"it was not diverse at all." But she told herself that was to be expected in this part of the country and she would keep an open mind.

The first week, her task was to complete a battery of online trainings on topics like cybersecurity, radiation protection, and diversity, equity, and inclusion, which involved reading educational content and taking quizzes. One day during that week, she entered a trailer with open workstations, sat at a table by a window, and began quietly working on a training.

As she was making her way through the material, she heard a voice coming from one of the offices at the end of the trailer. A man, apparently on the phone, was saying that he had rented out a room to an intern for the summer. When he had spoken with the intern on the phone a few months ago, the intern had sounded White. "But when he showed up to his house, the previous night, in a white pickup truck, there is this Black dude that comes out," Charlyne recalled. "And he's a bit confused, because he's wondering if the person has the wrong house." But he soon realized that the person didn't have the wrong house. "He's kind of taken aback, because he wasn't expecting the intern to be Black." He mentioned that the intern was a PhD student at UC Merced. And he said he had to come up with a way to get this guy to move out, because he didn't feel comfortable leaving him in the house on his own. Charlyne shut down her computer and just sat there, listening. "Listening until I couldn't listen anymore, and then I got out of the trailer."

Her assigned mentor, whom she liked and trusted, suggested that she report the incident to HR, though she'd not intended to before talking with him. She emailed HR, and she was moved to the trailer next door. The man, whom she never actually saw, was apparently moved somewhere else. When she was summoned to speak with HR staff, it seemed that they had already spoken with the employee. According to Charlyne, they told her, essentially, "Oh, you took it the wrong way." (Sarah Neumann, INL's PR person, wrote to me in an email, "INL takes allegations of discrimination very seriously and immediately launched an investigation. This included interviewing all the parties involved: [Charlyne], the overheard host of the intern, as well as the intern who was living with this particular host. Ultimately, the investigators concluded that the incident was taken out of context and this was communicated back to [Charlyne]."[2]) At the same time, while Charlyne thought her comments would remain private, word apparently spread quickly. "There were random people that were stopping by. 'Oh, I heard blah blah blah. . . .' I kind of felt like I was in a museum, and people knew exactly where to find me that I didn't know, that were just in my business."

About a week later, outside an administrative building in town, Charlyne saw a young Black man walking by. "And of course I turn around because I notice him because of course there are no Black people here." She dispensed with social niceties. "I immediately just went up to this man, and I was like, 'What kind of car you drive? What color is it? What school you go to?'" He gave her all the "right" answers—he drove a white pickup truck, and he was a student at UC Merced. She told him about the phone call she had overheard.

The whole experience cast a pall on her internship. Some people, she says, made comments suggesting that she was overly sensitive. She said the head of inclusion and diversity even convened a meeting where she brought up Charlyne's experience and suggested that there were different ways of responding—essentially that Charlyne had overreacted. (INL, through Neumann, denied this, but I spoke with three people who attended the meeting and backed up Charlyne's version of events.) "There were other times where I wouldn't say it was very explicit retaliation, but it was more a feeling of exclusion in discussions, or I was being dismissed. Or just ignored." Whether this treatment owed specifically to the incident or more generally to bias was hard to say.

The only silver lining was that she and the intern, Jordan, ended up dating.

WHILE CHARLYNE WAS IN GRADUATE SCHOOL, A HANDFUL OF PRO-NUCLEAR advocates were becoming attuned to the hierarchical and unwelcoming aspects of the nuclear world and were starting to try to address them. More generally, they were seeking to advance a progressive vision for nuclear power. In the summer of 2020, a new organization called Good Energy Collective launched. It was cofounded by Jessica Lovering, who had previously worked at the Breakthrough Institute; she had been hired to start up their nuclear program in 2012. She left that organization on good terms, but she wanted to pursue an agenda that would align nuclear with left-wing goals, building coalitions with progressives and environmentalists, which was clearly not a priority for the Breakthrough Institute.

Her new nonprofit's mission statement explained that nuclear energy would be needed for decarbonization, but "it's going to need to look different. We don't just mean new technologies—the entire industry needs to change from the ground up. Nuclear needs to grapple with the injustices of its history. We need a social science policy research agenda to match the engineering innovations. And most importantly, we need a diverse group of rising leaders who can partner with the climate movement to create an aggressive, coherent vision to address climate change and inequality."[3]

While, at the time, plenty of mainstream, moderate Democrats had come around to accepting nuclear, there was less enthusiasm, and more suspicion, further left. In an interview coinciding with the launch, Lovering's cofounder, Suzy Baker, told David Roberts, by this time at the news site Vox: "We took stock: 'Who are the progressives in the nuclear sphere? Oh, there's like five of us.' So it wasn't that hard to get everybody on the phone and ask, 'Do you want to do this?'" In the same interview, Lovering explained their focus on community engagement. "We want to develop tools and processes to help communities figure out what they want for their low-carbon energy future. And we want a process where they can decide they *don't* want nuclear. That's fine. We're not going around trying to sell advanced nuclear. We want equitable processes that people feel are fair and transparent so they can make their own choices."[4]

When I first heard about Good Energy Collective, I was intrigued. More than that—I was almost relieved. In my journey through the pro-nuclear world, I had sometimes been frustrated by what I saw as dogmatism, refusal to acknowledge almost any downsides to nuclear energy or the industry. (The main criticism from advocates of the industry was that it didn't do a sufficiently good job of communicating how great it was.) And I'd been disappointed by the frequent antagonism toward environmentalists that I'd observed—as well as perplexed, because all of this seemed strategically counterproductive. Good Energy Collective advanced the concept of "nuclear justice," which included bolstering diversity in the workplace, honoring community needs in future nuclear development, and redressing past harms. (As with nuclearism, this term had been introduced previously in the context of weapons, but as far as I can tell, Good Energy Collective was the first to use it in the context of energy.)

In March of 2022, I met Lovering for lunch at a café in a shopping center in Santa Barbara, where she lived. We sat at a table in the outdoor seating area, overlooking a vast parking lot. "I think part of what makes Good Energy Collective different is that we are really engaged and focused on the bad sides, the waste and the weapons history," she told me. "So we're definitely not defensive about it." If someone brings up a concern, they can say, "Yes, that is an issue! We can talk about that."

She echoed what I had been thinking—that this was not only a more tactful way of engaging with people but also more effective. "As opposed to something other groups do, in particular this category of people we call nuclear bros." They are focused on debunking and telling people they are irrational. "There's lots of studies showing the sort of debunking communication style entrenches people in their beliefs," she said.

The way she spoke about nuclear struck me as very pragmatic. Her group was focused on it not necessarily because they thought it was the best form of energy but because they believed it was one important piece of the puzzle, and that, compared with other vital energy sources, such as wind and solar, it needed more attention in terms of crafting a relevant progressive policy agenda. This ecumenical approach was reflected in their name: "We liked the idea of not having something nuclear-specific because that's not our end goal," she said. "The goal is to have good energy." And they didn't automatically support anything and everything nuclear. They were highly critical, for example, of fellow advocates who endorsed the infamous 2019 bill in Ohio that bailed out the economically struggling nuclear plants along with coal plants. And in a dispute over whether new uranium mining should take place near the Grand Canyon, they sided with the local tribes that opposed it.

I thought of a line from environmental journalist Amy Westervelt: "The problem is the power structure not the energy source."[5] I didn't entirely agree with the literal meaning of the sentence: different types of energy do have different virtues and drawbacks, of course. But it was true that some activists tended to get caught up in their allegiance to certain energy sources—whether renewables or nuclear—in what has been dubbed "energy tribalism."[6] I was starting to think that what mattered at least as much was the process for procuring our energy: the standards and safeguards for mining, whether uranium or copper (for solar) or lithium (for batteries); respectful communication and community engagement for siting infrastructure, whether a small modular reactor or a wind farm; and managing waste, whether spent nuclear fuel or solar panels at the end of their life, in a way that minimizes pollution and harm.

As Lovering put it over lunch, "Every energy source has waste streams. How can we manage all of them to reduce environmental impacts and impacts to public health? And how can we, when we're making choices about our energy system, sort of look at all the waste streams equally at the same time? Because there are waste streams from the manufacture of solar panels, the mining for renewables, the mining for batteries. But I don't want to make that point with someone who's very pro-renewables as, like, a gotcha . . . because it feels very, like, defensive, and like an industry talking point." She found it "more productive and more authentic" to say, "We gotta look at everything holistically. Because I do think renewables are great and really important and we need to scale them up a lot, but we also need to deal with their challenges the way we're trying to deal with nuclear's challenges."

After her summer internship ended, Charlyne did not particularly want to return to Idaho National Lab. She loved the technical aspects of working at the lab, which she likened to "a playground for engineers." But she never felt entirely comfortable or welcome there. Nevertheless, she did return, because it was the only place with the facilities she needed for her dissertation research. If she hadn't returned, she would have had to choose another project, adding at least a year to her studies.

After she finished her PhD, she also took a position as a postdoc at the lab to continue pursuing that line of research; she was already proficient in using the instruments there, and it was the only place that made sense for her career. But when she tried to collaborate with colleagues so she could work with new materials, her advances were rebuffed, limiting what she could learn and accomplish. She also thought it was imperative to change public perception and the regulatory environment. "Nuclear energy is not a technical problem, it's a political and social one," she said. So she began seeking other opportunities shortly after her postdoc began. She ended up taking a position as a nuclear energy analyst at the Breakthrough Institute, which had recently opened a new office in Washington, D.C.

In November of 2022, she moved from Idaho Falls to Maryland, where her parents were now living. Almost as soon as she started, she attended COP 27, the global climate conference, in Egypt. When she returned, she jumped into two of the priority areas the nuclear community was working on.

One was NRC reform. Some of the federal legislation that had passed in the 2010s had instructed the NRC to institute new policies to accommodate the variety of new reactors start-ups were pursuing. But this reform was not going smoothly. There was a philosophical question at the heart of regulatory debates: Should the goal be to eliminate as much risk as possible from nuclear reactors—the "precautionary principle"—or to take their benefits into account as well? The Breakthrough Institute was promoting the latter. Their point was that we cannot eliminate all risk, and when we try, the result is that nuclear plants do not get built, and we burn fossil fuels instead. Charlyne delved into analyzing the NRC's new alternative regulatory framework for advanced reactors.

Another area of excitement for nuclear advocates was the promise of coal-to-nuclear transitions. Because of increasingly stringent air pollution policies begun under the Obama administration, as well as economic factors such as cheap natural gas, dozens of coal-fired plants had closed in recent years. Coal capacity in the U.S. reached its peak in 2011, according to the

Institute for Energy Economics and Financial Analysis, but the country was expected to lose half this capacity by 2026.[7] This was good news for the climate and for air quality, but posed economic hardship for communities that lost a major source of employment. It also represented energy that needed to be replaced and, as the nuclear community saw it, a perfect opportunity to build new nuclear. First of all, these sites already had infrastructure and transmission lines that could be repurposed for nuclear plants. They also had energy workforces in need of new jobs. A nonprofit organization called Terra Praxis was working on facilitating these transitions, as was GAIN, the Obama-era program intended to help commercialize advanced reactors. The nuclear start-up TerraPower, backed by Bill Gates, was planning to build its first reactor in Wyoming, near a site where a coal-fired plant was scheduled to retire. Charlyne began working on designing a program that would streamline the permitting process for nuclear developers seeking to use these sites.

I met Charlyne in June 2023 at Cavallo Point, the idyllic resort in Marin County where the Breakthrough Institute held its annual conference. For several hours, we talked in her room, a suite with a sitting area overlooking the Golden Gate Bridge. She sat on a couch next to the window as the sun coming through the window slowly faded. She was wearing earrings in the shape of the symbol for the atom and, on her blazer, a pin with the green, gold, and black pattern of the Jamaican flag. The other major project she was working on was a "nuclear bootcamp" for Jamaican high school students. It was to be held over the summer at her alma mater. She wanted to educate young Jamaicans about what could be possible for them. "If I'd stayed in Jamaica, I would not have been able to become an engineer. It wouldn't have been an avenue for me."

The Breakthrough Institute, like Idaho National Lab, was also mostly White. The only other Black employee was another young Jamaican woman. But Charlyne found it more welcoming than the lab. "Yes, there are some microaggressions," she said. She said that one particular employee had, on multiple occasions, called her by the name of her Black colleague. "Nowhere's perfect, but it's tolerable, and I don't feel like I need to go and see a psychiatrist working at Breakthrough." (She would end up leaving, however, less than a year later, for a job at the Electric Power Research Institute, where she would use more of her engineering skills again.)

There is no simple way to categorize the politics of the current pro-nuclear community, or even of each of the groups and individuals that make it up. Charlyne doesn't consider herself particularly political; her goals are to expand nuclear energy and bring it to the Caribbean, but some of the obstacles she's encountered have had to do with personal politics in the field, and she's had to contend with them. The Breakthrough Institute, meanwhile, has in some ways seemed to move rightward, increasingly criticizing climate "catastrophism," but Ted Nordhaus has also published in the socialist magazine *Jacobin*. The nuclear world remains small and incestuous, with no shortage of internecine disputes. And in terms of getting its message out, this political heterogeneity may be a strength, with different advocates and groups reaching different constituencies. It also reinforces the impression made by the bipartisan legislation: that nuclear has been scrambling the usual partisan divides, offering a conspicuous exception to the rule of contemporary politics—even, arguably, challenging the coherence of our partisan categories.

CHAPTER TWENTY

DIABLO CANYON LIVES

At 12:43 a.m. on September 1, 2022, California state senator Bill Dodd, a Democrat representing Napa, took the floor in the senate chamber. Standing in front of the dais, the gruff-voiced lawmaker made the case to his colleagues that the life of Diablo—which he pronounced "Die-ablo"—should be extended.

His bill, SB 846, would authorize the extension of the plant by up to five years. It would make the plant eligible for newly available federal funding; the reason it had been rushed through, he said, was to meet the deadline to apply for that funding. "Members," he said, "Diablo Canyon supplies approximately seventeen percent of California's zero-carbon electricity, and more than eight percent of the state's total electricity. California's current clean-energy generation and storage system is not yet able to adequately backfill the energy production from Diablo Canyon Power Plant." He added, "Scrambling to buy costly and dirty out-of-state power isn't the answer and will hammer ratepayers and worsen climate change."

The bill also included off-ramps: for example, if the state were to build enough new renewable resources in the next few years to fully replace Diablo Canyon, then the original closure dates could be reinstated. "But we need to be proactive. We need to have the option available to keep the lights on and keep making progress towards net-zero emissions," Dodd concluded. "The risks to Californians are too great to chance an inadequate energy supply."

The day that had just come to a close, August 31, a heat wave had hit not only California but most of the western United States. That limited the state's ability to import out-of-state energy, because the other states needed it, too, for their air conditioners. At the same time, the ongoing drought had substantially reduced the amount of hydropower available. On August 31, Governor Newsom had proclaimed a State of Emergency, and CAISO had issued a flex alert, requesting that residents reduce their electricity consumption between 4:00 p.m. and 9:00 p.m. Californians were asked to "pre-cool" their homes before 4:00 p.m., set their thermostats to 78 degrees, and refrain from using their washing machines and dishwashers during that window.[1] It all prompted the question of what would happen without the nuclear plant to supply a significant chunk of the state's electricity.

The irony was that Democrats, who had historically opposed nuclear in general and Diablo Canyon specifically, were supporting this bill. Republicans had historically favored both, but the two senators who spoke in opposition were Republicans. They still supported Diablo, but they were protesting the way this bill had come about. "This crisis is created by a governor and a legislature that has no plan," said Senator Brian Dahle, a goateed lawmaker who represented a largely rural area in Northern California. "Sorry for yelling at ya, but for god's sake, let's have a plan." Senator Andreas Borgeas, who represented Fresno and bore a conspicuous resemblance to Conan O'Brien, said, "It's one o'clock in the morning, September 1, and we're exhausted. All of us want to get home, all of us want to get back to our districts tomorrow." He added, "Bad policy is made when legislators legislate from a position of fear. And I'm concerned that we are in a position not only of exhaustion but of fear." They saw this measure as bailing out the governor and the legislature from the consequences of poor decisions they had made in the past.

Over in the Assembly, the bill had just passed. In that body, there was more support from Republicans. It was introduced by Jordan Cunningham, the one who had sponsored the doomed constitutional amendment to bolster nuclear back in 2019. Jim Patterson, Republican of Fresno, said, "I don't think the governor had any choice. I don't think this body has any choice. If we want to be the Gray Davis of our generation, don't do anything." Devon Mathis, another Republican, defended the plant, and managed to get in a dig at Japan. "I know there's a lot of people that go, 'Oh, it's built on a fault line,'" he said. "This isn't Japan, this is California. The technology's a little better."

Democrats also spoke in favor. Eduardo Garcia, a Democrat from Coachella, noted that his district was bracing for temperatures of 115 degrees over the weekend. "We need air conditioning to take care of our seniors, kids at schools, just for the working families who work from sunup and sundown, who expect to go home and make sure that the lights are on and the AC is working." Chris Holden of Pasadena said, "I'm not a proponent of the Diablo Canyon Power Plant. But I am a proponent of keeping the lights on."

Ultimately, in the Assembly, it had passed 67–3, with 10 abstentions. When it came time to vote in the Senate, it passed 31–1, with 8 abstentions. On September 2, Newsom signed the bill into law.

After Newsom gave his bombshell *Los Angeles Times* interview in April 2022 and Feinstein's op-ed consolidated Democratic support, it seemed clear that the Newsom administration would pursue an extension. They drafted a piece of legislation, but waited until August to start pushing it. Bill Quirk, a Democratic assemblymember (now retired), told me that he had never been lobbied by the governor as intensively as he was on this bill. "His staff spoke with us multiple times in small groups." The chief of staff and experts would walk the lawmakers through the bill and why it was needed. "Never seen anything like it in my ten years," Quirk said. "Never."

As Newsom's staff shopped the bill around, they made some concessions. A climate-focused group of senators managed to include lavish funding—$1 billion—for renewables. The initial extension specified in the bill was for 10 years, but Senator John Laird, who represented San Luis Obispo and many anti-nuclear constituents, got that number reduced to five.

On September 2, the day Newsom signed the bill, Michael Shellenberger was a guest on *Decouple*, one of the podcasts that had emerged as part of the pro-nuclear media ecosystem, hosted by Chris Keefer, a Canadian physician. The episode was titled "Diablo Canyon Lives." "It's a massive event," said Shellenberger. "This is Diablo Canyon. This is a highly symbolic nuclear plant. . . . I feel very proud of, of having, you know, been a cofounder of this movement and a real early champion, in part because it was so lonely and painful in the beginning, and then to see so many people rushing into it. You know, as they say, failure's an orphan and success has many fathers. And so I think we're seeing a lot of people taking credit for what's going on today."[2]

I started paying attention to the campaign to save Diablo Canyon in 2019. I thought Mothers for Nuclear and the other advocates might succeed in changing the larger conversation about nuclear, but I did not expect them to succeed in their crusade to rescue the plant—and I don't think they did, either. How did the politician who oversaw the original deal end up becoming the one to overturn it? Who does deserve credit (or, depending on your perspective, blame) for the reversal?

The general consensus held that it was essentially a matter of avoiding blackouts. Newsom did not want to be the next Gray Davis, and all the other elected officials knew they would be faulted, too, if blackouts occurred on their watch. But of course, there's much more to the story than that.

That was certainly one factor. Back when I first learned about the deal to close the plant, and the arguments that doing so would harm the climate, I considered the situation from a politician's perspective. Climate change is such a diffuse and gradually unfolding problem; no single elected official is going to be blamed for it, which is part of why it's been so hard to address. If a nuclear meltdown happened while you were in office, though—because you had allowed an aging plant to continue operating—accountability would be harder to dodge. I wondered if that reasoning entered into the calculation for Newsom and the other state officials. Now the calculus had changed again. Blackouts were less scary than a nuclear accident, but seemed more likely to occur; and apparently averting them had become the top priority. "Certainly that was the reason—the governor would never have supported it otherwise, I guarantee you that," Quirk told me.

When I asked the pro-nuclear advocates themselves, they gave themselves a moderate amount of credit. "I think that the role we could have played was to give him coverage," Isabelle Boemeke told me. The open letters, the social media advocacy, the rallies, and the public comments at myriad hearings and meetings to counter the anti-nuclear comments—it all showed that there was a contingent in favor of extension. "I don't think it would have happened if California didn't have reliability issues," Isabelle said. "But I think it did provide a good coverage for him." (Although she was not forthcoming about any behind-the-scenes efforts, her fellow advocates told me that her connections had tipped her off when a strong pro-nuclear showing was needed at a particular hearing, and that she had had conversations with Newsom's donors.) Gene Nelson also told me that a generous anonymous donation had enabled him to hire a high-priced lobbyist who had a close contact in the governor's office.

Also noteworthy was the absence of widespread opposition. Groups like Mothers for Peace and Friends of the Earth were furious, of course, but the response was hardly on the same scale as the protests in the '70s and '80s. The general public sentiment seemed to be a shrug.

Some of the influence of the pro-nuclear movement was direct: those advocacy efforts, the funding and execution of the Stanford/MIT study. Some of it was less direct, through their work to sway public opinion on nuclear more generally, as well as the effective lobbying for new federal laws. After all, the availability of federal funding was explicitly linked to the extension of Diablo Canyon: Newsom was motivated in part by the Civil Nuclear Credit Program, included in the Bipartisan Infrastructure Law to preserve existing nuclear plants. While the passage of SB 846 was not the end of the story—as always in California, there would be reviews from multiple agencies pending, in addition to the application to the NRC—it now seemed all but certain that the plant would stay open for at least an additional five years.

The episode ultimately underscored the importance of reliable electricity in today's world. It is nonnegotiable, and politicians who can't guarantee it will pay a price. This in itself arguably shows the value of nuclear, for its ability to provide steady, baseload, low-carbon power. Elected officials, and even some mainstream environmentalists, were also aware that if California did stumble—if the state proved unable to keep the lights on—that failure could be a huge setback for the clean-energy transition, as naysayers could point to California blackouts as a reason to stick with fossil fuels. Siva Gunda, vice chair of the California Energy Commission, told me that Newsom had realized that "having reliability issues in the clean-energy transition, it's not an acceptable thing. And you need to protect the entirety of the vision . . . it was like all hands on deck, and everything is on the table."

The next month, in October, I met Heather and Kristin for a hike on the Point Buchon Trail, one of the two trails with public access on PG&E property, this one at the northern end of the Diablo Canyon Lands. The trail is a little over three miles, and culminates in an overlook called Windy Point, from which you can see the nuclear plant in the distance. Other than the access road, this vista is one of only a handful of places where the plant is visible from land.

As we walked, I asked how they felt about the passage of SB 846. They were less triumphalist than Shellenberger and some other advocates had been publicly. Specifically, they were frustrated by the limit of the extension to five years. It was still being framed as a bridge to an all-renewables future—the bridge was just getting stretched out a little.

"It's, like, the most inefficient way to extend the plant life," said Kristin. PG&E would be in a position to do only minimal upgrades and updates, rather than long-term investments. "It's so silly."

"We predict that we're going to operate for at least twenty more years," Heather added. "It makes so much sense. It's so obvious to so many people for so many reasons. It's good for absolutely everyone."

Indeed, NRC licenses are typically for 20 years. Regardless of the language of the legislation, PG&E would be applying for a new 20-year license. If the electricity continued to be needed, the legislature would likely end up in the same position in five years—perhaps passing another stopgap measure.

It was a cloudy day with some blue sky peeking through the clouds. The trail was right along the coastline, with the same landscape of coastal shrub and chaparral I'd seen when driving out to the plant several years before. I noticed patches of small yellow wildflowers, which Kristin and Heather told me were California poppies. On the rocks and arches out in the sea, we saw black birds with long, bright orange beaks: black oyster-catchers. Because they are threatened, there's a state program to monitor them. Heather and Kristin used to come as volunteers with their kids and count them.

Notwithstanding their frustration about the specifics of the law, they were feeling optimistic in general. "I think the Diablo Canyon decision, the political signals that that sent, we're already seeing the ripple effects from that, and we're gonna keep seeing it, which is exciting," said Kristin. "So Heather and I are thinking, 'Okay, so where else needs a little more of a push?' Now that this permission has been given, if you will, by this powerful Democratic leader—he changed his mind."

"We feel like now's a really good time to capitalize on all this great momentum for nuclear, and, like, really push the envelope," said Heather. "Not just saving Diablo Canyon in California. That's like maintaining the status quo. That's how much clean energy we have. We're not going back-ward. But we need more. We need new nuclear. We need to repeal our nuclear moratorium in our state. And we need to start pushing for advanced nuclear. So now's the time, right?"

Finally we arrived at the lookout point, where we could see the plant to the south. There was an informational sign, which read, "Delivering Clean Energy to California," alongside a photo of workers in hard hats. We sat down to rest on a bench. The salt wind whipped our hair. As we were rest-ing, a young couple arrived.

"This is so cool!" said the young woman.

"Have you guys seen it before?" Heather asked.

"No!" She introduced herself as Cayla, a Cal Poly student. Heather and Kristin told her they worked at Diablo Canyon and answered a few of her questions. "I feel like it's not talked about enough," Cayla said. For a few moments, she and her boyfriend stood gazing out at the plant, their arms around each other.

WHEN I TALKED WITH JANE SWANSON, OF MOTHERS FOR PEACE, OF COURSE I heard a very different response. As we sat at her dining table, surrounded by the books about composers and classical music, I asked if she was shocked by the reversal. "Yeah! Absolutely," she said. "I thought, 'They can't! They signed that Joint Proposal.'"

Mothers for Peace was still a local force, although it had dwindled over the years. A few members had moved away; more had died. One member regularly updated the timeline on their website—which listed all of the significant events related to the group and Diablo Canyon, dating back to the proposal to build the plant in 1963. This member was 96, Jane told me. "Her hands are so crippled with arthritis, I don't know how she does it." They now had Zoom meetings about once every three weeks, and usually five or six people, other than Jane, attended. They had hired a publicist, and they had a minimal social media presence, but they couldn't compete with the likes of Isabelle Boemeke and Michael Shellenberger.

Jane was aware of this. The pro-nuclear side had long seen themselves as underdogs, but so did the anti-nuclear side. "Look where the money is," she said. "Look where the power is. It's not really with Mothers for Peace and our allies."

She was also, like her adversaries, acutely aware of the symbolism of Diablo Canyon. The supporters of Diablo, she said, "want the nuclear renaissance. That's the big picture. If they can succeed—all those bad people—if together they can keep Diablo going for another twenty years, it's like, look, they restarted this one that's sitting on thirteen active earthquake faults so, like, let's restart ours. So it's like a big symbol."

I asked if it had been hard, living near the plant all these years and fighting it, only to see the end in sight and then have to fight it again.

"It's a burden. It's a burden. . . . There are times when I get really discouraged. I don't think I've ever been clinically depressed, but there have been times when I felt—a little bit depressed. And that's why it's so wonderful to be part of a group, because your friends can help you with perspective. And sometimes I can help them with perspective. Because there have been times when I have been so discouraged and thinking, 'What kind of a place is this world?'"

But, she said, she had landed on a coping mechanism. "When I get down, I have adopted a philosophy that helps me." At these times, she said, she thinks, "'Okay, wait a minute, go back to your root philosophy,' which is, no life-form is permanent. It's the truth. No matter what the environmental movement does. No matter what actions we take to slow down climate change. We can slow down these horrible things, but we're not gonna stop 'em. So in the end"—she paused—"it's gonna be cockroaches."

She went on, "If I actually thought that if people worked really hard, they could stop climate change, they could stop nuclear power, they could stop war, I mean if I thought these things were possible, then I would be so frustrated and depressed. But just accept the reality that it's the cockroaches, and you just do the best you can, that's all you can do." (She later clarified, "They're very resistant to radiation. I didn't just pick cockroaches out of the air."[3])

She wasn't about to stop fighting Diablo. There were several avenues Mothers for Peace could pursue to thwart the relicensing. For example, because the time frame for applying for license renewal was so compressed, the plant would need an exemption from the Nuclear Regulatory Commission. Mothers for Peace would challenge the exemption in court. "Although one way or another, the cockroaches are going to win," Jane said, "I want it to be a long time before they win."

Some months later, I visited the home of Mona Tucker, the chair of the YTT Northern Chumash Tribal Council. Scott Lathrop was there, as was Wendy Lucas, vice president of the nonprofit. They met me outside Mona's ranch-style house in Arroyo Grande, another city in San Luis Obispo County, to the southeast of Diablo Canyon. Inside, we went to the kitchen, where Mona served slices of apple bread her husband had baked for us. We took our plates to a table in the living/dining room. Framed family photos, as well as a picture of a Cooper's hawk, adorned the wall. Artisanal birdhouses, which Mona's husband had built with salvaged wood, were propped up on bookshelves.

Mona, the leader, sat at the head of the table. She was firm, on message, and cautious about talking to the media—i.e., me. She started off with this: "We want people to know who we are, where we're from, and what we want. Who we are, we're the Indigenous people of San Luis Obispo County and region. This is well documented."

Mona continued: "And where we're from, we're from here. We're from the San Luis Obispo county and region. And the three of us, plus most members of our tribe, are linked directly to the villages that are on the Pecho Coast. And what we want, we want returned to us what was taken from us in the seventeen hundreds. And it was taken violently, without our permission, without agreement, without compensation. There was no hesitation to enslave us, to kill us, or to bring about ideologies that resulted in our death, diseases, et cetera. So we want our land back."

One might have expected them to be infuriated about the reversal of the decision to shut down the plant. But Mona presented their attitude more pragmatically. "We don't have a position, other than, if the plant stays open, we want our land back. If the plant closes, we want our land back. It changed our strategy. It changed a lot of other things. But as far as our goal, it never changed." When I asked for her views on nuclear energy generally, she said, "We're concerned about climate crisis and we're concerned about power needs, but our main focus is getting our land back. As we've said in the past, if the plant stays open, we want our land back. If the plant closes, we want our land back."

If the tribe owned the land, they could lease Parcel P, the area with the industrial infrastructure, to PG&E, and start to recoup some of the massive value that they were never able to benefit from before. "The use of our property that was stolen from us has generated more money than I can even count, and we've never received a penny," Mona said.

As for the land surrounding the plant, the other parcels, they think those should be returned to them whether the plant is running or not. They insisted that their goal is simply conservation. "It's not a casino," said Mona. "And please publish that quote. It's not for a casino. It's not for development. It's not for a resort. It's not to break it up into pieces for twenty-million-dollar ranchettes, or thirty million. It's not that at all. It's for conservation of a very special part of the coast of California on the western edge of the United States. It's our homeland."

They envision working with the Land Conservancy, one of the organizations they've partnered with, to preserve the land, perhaps adding a couple of caretakers' quarters. They'd like to have an education center for the public to learn about their tribe. They also say they would open up public access somewhat, beyond the two trails that are currently accessible. But the tribe members emphasized that caution would be required to protect their cultural inheritance. The land has been studied, but there may be more burial grounds, artifacts, and other remnants of their ancestors' lives that have not yet been discovered.

The tribe has had conversations with PG&E, and they believe they have a good relationship with the utility. They were also able to work with the governor's staff on SB 846. The final text of the bill stipulated that "all relevant state agencies and the operator of the Diablo Canyon powerplant should consult and work collaboratively with local California Native American tribes . . . to consider tribal access, use, conservation, and comanagement of the Diablo Canyon powerplant lands and to work cooperatively with California Native American tribes that are interested in acquiring such lands." They also won a proviso that "existing efforts to transfer lands . . . shall not be impeded by the extension of the Diablo Canyon powerplant." At the time of my visit, however, they seemed to be at an impasse. Negotiations had stalled. It seemed that, perhaps not surprisingly, the state was more interested in paying lip service to the idea of landback than in actually making it happen.

Scott sometimes reflected on how the plant had come to occupy his ancestors' land. "I always find it amazing that the primary reason the plant is where it is, is because of a deal that PG&E cut with the Sierra Club about relocating it from the dunes to Diablo Canyon," he said. And then, of course, many Sierra Club members had come to fiercely oppose that deal, as they began to appreciate the majesty of the site. The irony was that because of the nuclear plant, thousands of acres of adjacent territory were preserved in pristine condition. The counterfactual is unknowable—maybe it would have been set aside as a state park—but it's possible that it would have been developed and altered beyond recognition.

The tribe's vision was stirring: beautiful open space conserved, comanaged by the tribe that lost it hundreds of years ago. And on the industrial parcel, cleantech innovation and an economic engine for the county.

"I always tell the story, many, many moons ago, you had a person and they would maybe take down a deer," Scott said. "Well, they wouldn't just kill a deer and take the hindquarters and go and leave everything to waste." Speaking for himself, he told me he hopes that whenever Diablo Canyon is decommissioned—whether that happens in 5 years or 35—it might be replaced with next-generation reactors. "Keep that plant operational as long as possible, and plan for conversion to new nuclear, new advanced nuclear," he said. "In other words, take that site that has been created as a resource, and maximize the use of that resource. Take that deer, and maximize the use of that deer."

As we wrapped up our conversation at Mona's house, I asked if they were optimistic about recovering the land of their ancestors.

"I have faith," said Mona. "It's our land to start with—"

"Yeah, it's coming back sooner or later," Scott added.

"—and it's coming back."

EPILOGUE

When I started working on this book, I had two big questions: Is nuclear power actually good? And will these wacky pro-nuclear activists actually get anywhere? Over the course of writing it, the second question has morphed into a new one: *How* have these wacky pro-nuclear activists gotten so far so fast?

The movement has been remarkably successful, by a number of measures. In August 2023, the Pew Research Center reported, "A majority of Americans (57%) say they favor more nuclear power plants to generate electricity in the country, up from 43% who said this in 2020."[1] In addition to the new federal laws, states throughout the country have begun changing their laws to allow or encourage new reactors. Starting in 2016, Wisconsin, Kentucky, Montana, Connecticut, West Virginia, and Illinois all repealed or modified their restrictions on building new nuclear reactors.[2] Several nuclear plants—most saliently, of course, Diablo Canyon—have been rescued from planned closures. And at COP 28, the climate conference held in Dubai in 2023, more than 20 countries, led by the U.S., pledged to triple global nuclear capacity by 2050.

The reasons for the movement's success are hard to distinguish from the conditions that gave rise to it in the first place. Growing concern about climate change made the attributes of nuclear energy—a steady, low-carbon source of electricity—more appealing. The trauma of living with fear of nuclear weapons faded from memory for some, and was never an issue for others. During this time, well-known scientists like James Hansen, Steven Chu, and Carolyn Porco were advocating for nuclear energy as a solution to climate change, and international bodies such as the IPCC and the International Energy Agency were presenting it as an important option, too.[3]

And all of this happened at a time when faith in science and scientists was ascendant in progressive circles. Already, in the context of the climate debate, scientists had a great deal of credibility in those milieus. Then, during the COVID-19 pandemic, attitudes toward expertise and science polarized even more, as Democrats solidified their deference to bodies such as the CDC and the WHO, while Republicans held these institutions increasingly in contempt.

Advocates could also point to several studies from reputable sources indicating that the dangers of nuclear had been overblown. The WHO reported that as of 2005, almost 20 years after the accident, fewer than 50 people had died as a result of radiation exposure from the Chernobyl accident—which was by far the worst nuclear accident in history. Those who died were nearly all rescue workers who were exposed acutely. The report predicted that that number could eventually climb to 4,000, as older people eventually died of cancer that could have been related to the exposure.[4] While this death toll is terrible (and the accident had other devastating impacts as well), it hardly sounds like the apocalyptic catastrophe it was feared to be at the time.

To be sure, other estimates—from Greenpeace and MIT historian of science Kate Brown, for example—are far higher. But these discrepancies pose a conundrum: When should we trust the official authorities, and when should we distrust them? When is a deep suspicion of expertise and authority warranted, and when is it simply conspiracy-theorizing? In the '60s, environmentalists and the left were more inclined to question the official line. In the 2020s, it has become the right wing that is more likely to dismiss it.

There were also new studies with remarkably reassuring findings about what had been one of the scariest aspects of radiation—the prospect of heritable genetic effects. In 2021, an authoritative study of the children of Chernobyl survivors, including cleanup workers, was published in the leading journal *Science*. The study found no evidence of transgenerational effects.[5]

It seemed, more and more, that science and expertise were on the side of nuclear power. It was not unanimous, but there were only a handful of high-profile dissenters who were also experts—Stanford's Mark Z. Jacobson, Amory Lovins—that the anti side could point to. There were no letters signed by dozens of scientists and experts to urge Governor Newsom to shut the plant down. So progressives, generally inclined to both worry about climate change and trust scientists, gradually became more open to nuclear.

One might have expected conservatives, then, to go in the other direction and become increasingly anti-nuclear. But that hasn't happened, for, I would guess, a couple of reasons. One is that attitudes toward nuclear are still quite mixed on the left, and another is that the shift in attitudes isn't necessarily widely appreciated. It certainly has not (yet) gotten to the point that support for nuclear is branded as a "woke" position.

In the end, though, I would venture to say that perhaps the most important reason for changing public opinion on nuclear energy is also the most banal: it is no longer new. As I know from social science research and personal experience, unfamiliarity is a huge part of what makes a threat disturbing. When nuclear was new, it seemed strange, and by definition, the long-term effects were unknown. In the foreword to Gofman and Tamplin's anti-nuclear treatise *Poisoned Power*, Senator Mike Gravel wrote, "If there were serious nuclear pollution, most healthy people might have to spend much of their lives caring for sick people."[6] This seemed like a real possibility if nuclear power plants were to become widespread, and who was to say it wasn't? Today, the country that derives the highest proportion of its electricity from nuclear fission—64.8 percent as of 2023—is France, which built out its nuclear fleet starting in the '70s.[7] We don't exactly think of France as a radioactive dystopia of illness and death.

WHO, IN FACT, IS THE UNDERDOG IN THIS FIGHT? THE PRO-NUCLEAR MOVEment liked to see itself as scrappy, grassroots, the first time "civil society" got behind nuclear. But it's always been better-funded, and more intertwined with industry, than that image would suggest. From its inception, as we've seen, it attracted support from wealthy donors—the Pritzker heirs, venture capitalists such as Ray Rothrock, and Silicon Valley millionaires such as Ross Koningstein. Several prominent activists I've heard described as "grassroots"—Eric Meyer, Paris Ortiz-Wines, Mark Nelson (of Radiant Energy Group)—have received some portion of their funding from companies in the nuclear field.

Based on my interactions with them, I believe these groups came to their views on nuclear sincerely, then sought out funding, and found that—since nuclear wasn't very popular, precisely what they were trying to change—there weren't a lot of options for philanthropic funding other than industry. They aren't AstroTurf—the term for when industry creates fake grassroots groups—but they are fertilized by industry.

When I asked Eric about this, he said, "In a sentence, because the work we're doing is"—he paused—"so *fucking* important—throw that bomb in there—we will take money from anyone that agrees with us on what we need to do." He went on, "Having strings attached isn't allowed. There needs to be a uniform kind of understanding of the vision for the money and for what success looks like as well." (He acknowledged that when the news had come out about the corruption at FirstEnergy, the Ohio company that had given him some funding to knock on doors, "that makes all of us look like assholes.") He and Paris both told me that they were trying to build up other funding sources, like individual monthly donations from supporters.

Nuclear advocates also correctly point out that mainstream green groups have taken money, in certain high-profile cases, from fossil fuel interests. David Brower relied on a donation from an oilman to start Friends of the Earth[8]; from 2007 to 2010, the Sierra Club took more than $25 million from the natural gas industry to fund its Beyond Coal program.[9] (Why would fossil fuel companies fund environmentalists? Possibly to expiate their sins or, according to the more conspiracy-minded view, because renewables don't pose a real threat to fossil fuels, so the success of anti-nuclear, pro-renewables groups would benefit them. In the case of the Sierra Club, the motive seems fairly straightforward: natural gas is a competitor with coal.) I think it's important to follow the money, to be transparent about it, but I also think all money is dirty if you follow it long enough. The major green groups are not struggling financially. As of 2022, the Sierra Club had more than $96 million in net assets[10]; NRDC had more than $475 million.[11] Ultimately, I really can't describe either side as the little guy.

Sometimes, when I learned details about the campaign to save Diablo Canyon, they seemed a little sketchy: the back-channel communications, the lobbying, Gene Nelson's anonymous donor. Based on my reporting, the pro-Diablo side seems to have been savvier and better organized, with more connections in state government, than their opponents, who seem to have been caught off guard. It was not a pure and angelic campaign, but as far as I can tell, it was ultimately typical politics. While the pro-nuclear movement was never quite the scrappy underdog it claimed to be, nor is it merely the mercenary outgrowth of industry that some of its adversaries suppose. If you think nuclear is evil, their activities seem sinister. If you think nuclear is our savior, then they seem like passionate activists carrying out noble work.

WHICH BRINGS US BACK TO MY FIRST QUESTION—IS NUCLEAR GOOD? I don't have a simple answer. But I think it depends in part on the answer to another question: Is energy use good?

The utopia of an environmentalist like David Brower was a world in which humans pulled back, reduced their numbers, curbed their energy use, and shrank their impact on the Earth. And while this vision might have entailed some sacrifice, its proponents thought that on balance, people would be happier in a world less dominated by consumerism and competition; after all, what made *them* happiest was hiking through sublime wilderness. What's more, sacrifice of some comforts and conveniences for the greater good was seen as both a spiritual and planetary boon.

Another utopia is a world imagined in the early 20th century, the early days of radiation science—one in which abundant energy, once we figured out how to unlock it, would be a liberating force. Before the birth of that new science, "we have adhered to the view that the struggle for existence is a permanent and necessary condition of life," said Frederick Soddy, a British Nobel laureate, in a lecture series in 1908. "To-day it appears as though it may well be but a passing phase[.]" In place of this struggle, he foresaw an "unlimited ascent of man to knowledge, and through knowledge to physical power and dominion over Nature."[12]

At heart, I have always been more drawn to Brower's vision, but reason tells me that we are not going to return to a pastoral lifestyle. Electricity use is projected to soar in the coming decades. Hearing the stories of people like Isabelle Boemeke and Charlyne Smith, and reading the ecomodernist literature, also helps me to appreciate the value of reliable electricity that I have admittedly taken for granted—that we probably all take for granted, once it becomes part of the fabric of our lives.

To be a bit reductive, I've come around to thinking: nuclear isn't as bad as I thought, and renewables aren't as good as I thought.

According to the Union of Concerned Scientists, "a huge increase in solar power production will require a surge in the mining of raw materials. There are myriad problems that exist with the mining of silicon, silver, aluminum, and copper needed to make solar panels."[13] Converting silicon to polysilicon, a step in the manufacturing process, requires very high temperatures; it is done mostly in China, and is largely powered by coal. Aluminum is sourced from bauxite, the mining of which is land-intensive and often intrudes on Indigenous land. Wind turbines, meanwhile, consist partly of steel, and making steel is currently very carbon-intensive.

The limitations of wind and solar (such as intermittency) and the ecological costs (such as mining and land use) can both be mitigated, but alleviating one problem often exacerbates another. For example, lithium-ion batteries can provide backup, but they have their own environmental and human rights implications. Another strategy to smooth out the supply would be to overbuild renewables and connect them over a large region. The sun is always shining somewhere, just as the wind is always blowing somewhere; at 3:00 p.m., California could export its solar energy eastward, to places where dusk is falling. But overbuilding, of course, requires more mining and steel, and more land use, while the integration over large areas requires an extensive network of transmission lines.

Renewables once seemed abstract: a pure, benign way to harness the power of the elements. But as they begin to scale up, the reality becomes harder to ignore: that they are material infrastructure that requires mining and occupies land. In February of 2023, Greta Thunberg joined a protest to demand the demolition of a wind farm in Norway. The 151 turbines, each rising to 285 feet, had been erected in 2020 on the land of Indigenous Sami reindeer herders. According to the herders, the gigantic turbines, with their loud whirring, frightened the reindeer, disrupting the herders' way of life. Thunberg, wearing a knitted hat, cream-colored mittens, and a black puffer jacket, sat down with other young protesters to block the entrance to Norway's Ministry of Finance. "I'm here to support the struggle for human rights and Indigenous rights," she said. "We want the windmills taken down." Then she was arrested, carried by two cops, one in front carrying her legs, the other holding her under her arms, as her body hung limply in a V shape.[14]

The idea of Greta Thunberg protesting a wind farm might sound like the president of the NRA protesting a gun show. But she is not the only environmentalist to object to a renewable energy project. The Audubon Society and local chapters of the Sunrise Movement and the Sierra Club are among the groups that have done so in specific cases. Native American tribes and their allies have also challenged plans to mine on their territory for raw materials needed for the clean-energy transition: in Nevada, the Shoshone and Paiute Tribes have been protesting a proposed lithium mine on their sacred land, while in Arizona, the San Carlos Apache Tribe objects to siting a copper mine on their ceremonial grounds.[15] The terms "green extractivism" and "green colonialism," which may have once sounded like oxymorons, have become rallying cries.

My point is not to bash renewables, which I believe have a critical role to play, but to honestly grapple with the problems that will attend the vast expansion of any industry. And nuclear has major challenges of its own: the upfront cost, the time plants take to construct, and the risks, which, while they may have been blown out of proportion, are real. If nuclear is to scale up throughout the globe, there will be serious questions about how to manage those risks responsibly and effectively. But its advantages—its energy density and its ability to provide steady power—cannot be lightly dismissed.

In the months before this book went to press, it was hard to keep up with all the news in the energy realm. The price of solar continued to plummet, installations continued to accelerate, and battery deployment rapidly increased. Geothermal energy also seemed to be on the cusp of a breakthrough, potentially providing another important source of clean firm power. Interest in nuclear remained strong, too—thanks in part to the rise of AI, an energy glutton. In September 2024, Constellation Energy announced an unprecedented move: it planned to reopen a reactor that had closed five years earlier. Remarkably, the reactor in question happened to be Three Mile Island's Unit 1 (not the unit where the accident occurred). The plan hinged on a deal with Microsoft, which had agreed to buy the plant's electricity for 20 years to power its data centers. ("Microsoft restarting Three Mile Island was not on my 2024 bingo card," one nuclear policy analyst told me.)

Given the pace of change, I would not presume to predict what the energy landscape will look like even a few years from now. To some observers, the developments in renewables suggested that before long, nuclear would not be needed. But writing it off strikes me as premature. In geographic regions with long dark winters, for example, and for industrial applications that require heat, fission could be the most viable low-carbon option. There are countless considerations that go into energy choices. My hope is that we can move toward a world in which these choices—including choices about how much energy to use, and for what purposes—are made more democratically, with greater understanding of the various trade-offs.

Despite, or because of, the nearly religious feelings expressed one way or the other about nuclear power by many of the people I've talked to, I've ended up in between. I'm not a nuclearist nor an anti. I've come to a view that is certainly not original; it's the one I heard that I found most convincing. Nuclear is not necessarily this mystical, sui generis phenomenon. Like any energy source, it has upsides and downsides. It's neither a panacea nor an abomination. In the course of writing this book, sometimes I've thought that we should use it only for what we can't do with renewables. Other times, I've thought it would make sense to derive the bulk of our electricity from nuclear, as in France. But I've never thought we should abandon it altogether.

In the two decades since the New Apollo Project, the ideas animating it—an emphasis on jobs as well as climate, on unleashing ingenuity rather than demanding sacrifice—seem to have gained traction. The Inflation Reduction Act (IRA)—passed in 2022 and hailed, despite its misleading name, as the most significant piece of climate legislation in U.S. history—constituted a massive investment in the clean-energy transition, with little in the way of regulations, restrictions, or taxes on fossil fuels. It was straight out of the "Death of Environmentalism" playbook; it was the industrial policy that Nordhaus and Shellenberger and their allies had urged.

Influential thinkers on the spectrum from center-left to far-left, such as Matt Yglesias, Ezra Klein, and Bhaskar Sunkara (founder of the socialist magazine *Jacobin*), promote an "abundance" agenda that emphasizes investment and building. They express concern that well-intentioned regulations, whatever their past utility, are now thwarting the development that needs to happen to solve our current problems, including climate change. Klein has coined the terms "supply-side progressivism" and "a liberalism that builds" to describe this philosophy.[16]

To be sure, this camp has yet to prove that their approach can solve our problems. And the older, more familiar strain of environmentalism persists. In June 2023, the new climate publication *Heatmap* published a piece titled, "The Climate Coalition Is Threatening to Split Apart." One of the points of contention between the factions involved what is known as "permitting reform." The build-out of renewables was facing serious hurdles because permitting for projects was so bogged down in red tape—red tape whose purpose was largely to ensure environmental protection. Some climate activists, whom the author of the piece, Josh Lappen, dubbed "green growthers," were in favor of reforms to speed up the process, but others—the more traditional environmentalists—worried that the same reforms would risk environmental harm. "For some green growthers, deregulation is a necessary precondition to decarbonization, and since many also believe that clean energy will—with the IRA's help—outcompete fossil fuels, they see fewer risks to reforming environmental law than the environmental movement does," Lappen wrote. "Permitting reform is threatening the national climate coalition because it cuts to the heart of a longstanding philosophical disagreement about what it will take to actually achieve decarbonization."[17]

In other words, as we've seen dating back to the early days of the Sierra Club, the traditional environmental movement is focused on stopping what's bad. The new "green growthers" are focused on encouraging, building, what's good. And while it's possible to be a green growther in favor of 100 percent renewables, most of them favor nuclear as well. It fits with their belief in technological progress, in human ingenuity—their guarded optimism that if we really put our minds to it, people can prosper while allowing nature to thrive, too.

One of the most committed members of the older camp is someone I know well: my mother. Like Jane Swanson and David Brower, she never sold out. She still shops at Goodwill and flea markets. And she is as dedicated an activist as she was in the '60s. As a teenager, she had planned to become a nun, but when she had her political awakening in college, she channeled all of her religious fervor into ending the Vietnam War. Later, she poured it into the fight against climate change.

She is no fan of nuclear power, but she has, for the past couple of decades, focused her efforts on stopping fossil fuel projects. She has protested dozens of times at the headquarters of an obscure federal agency with outsized power, the Federal Energy Regulatory Commission, which she and her comrades believe rubber-stamps fossil fuel projects. She has traveled to Minnesota to protest the Line 3 tar sands pipeline and to West Virginia to protest the Mountain Valley natural gas pipeline. I've lost count of how many times she's been arrested for civil disobedience. (When my daughter was seven or eight, she surprised her friends by mentioning that her grandmother had gotten arrested again.)

My mother and I are very different people. I suspect it's partly generational—she came of age in the '60s, after all, and I came of age in the '90s, at the tail end of Gen X. She has always been certain in her convictions and willing to act on them, whereas I am always more unsure and ambivalent, prone to stand back and analyze rather than jump in and do something. Or maybe it's just that I'm lazy. My generation, after all, has always been known as a bunch of slackers.

In any case, I see in my mother the kind of approach to the world that I see in Jane Swanson, the philosophy Jane explained to me, the philosophy of cockroaches (although my mom wouldn't put it in the same way). Jane's framing is, in a sense, certainly pessimistic. The premise is that everything good will disappear eventually, and that the best we can do is to slow this inexorable loss.

Yet optimism isn't necessarily right, and pessimism is not necessarily wrong. I can't help admiring them for doing all they can to slow the loss. I do hope there's a way to do both, to stop the bad and build the good. But I recognize that there are inherent conflicts that will never be resolved, such as, most saliently, fierce disagreements about in which category nuclear energy belongs.

On one level, Jane is clearly correct—nothing is permanent, and over time, everything is lost. This, after all, is the essential insight of Buddhism. But I have an alternative to the cockroach philosophy. I have the bird philosophy.

When I was a kid, I don't think they had figured out yet that birds had descended from dinosaurs. I picked up on this in bits and pieces in recent years. And it seeped into how I think about the world and the future. I love birds. I am not a birder—I am not obsessive or knowledgeable; I have no clue what most of them are called. But seeing their flight and hearing their song is one of my most reliable sources of joy, as it is for many others. Very little makes me happier than the thrum and blur of a hummingbird's wings. And birds, apparently, emerged from a truly apocalyptic event: the asteroid that wiped out the dinosaurs and most other forms of life on Earth.

This doesn't mean that I think apocalypse is fine, or that an asteroid strike is the same as human-caused environmental problems. But it does give me a new perspective, and some comfort, when contemplating what the future might hold.

I floated this alternative lens on the world by Jane. She said she would think about it.

ACKNOWLEDGMENTS

MY FIRST THANKS GO TO THE PEOPLE FEATURED IN THIS BOOK, WHO SPENT their time (in some cases quite a lot of time) sharing their knowledge and their views with me and telling me about their lives. I feel so fortunate that conversations with interesting people are part of my job.

I am also extremely fortunate to have David Halpern as my agent. David shepherded my proposal through multiple iterations, he always has my back, and he's fun to talk to. Without him, this book would certainly not exist.

Madeline Jones is an outstanding editor, but she's so nice about it that you hardly notice. Without her trenchant questions, and her sense of language and structure, this book would certainly be worse.

Josh Rothman, who edited the article on which the book is based, also has a deceptively light-touch approach. Thank you to Josh, and to Jessica Winter, for the opportunity to write that article. Alexa McMahon skillfully edited another article, on nuclear waste, parts of which made their way into this book.

Many thanks to Drs. Todd Allen, John Johnson, and Yoon Chang for generously lending their expertise and reviewing sections of the manuscript. Also, thanks to Brad Scriber for help with fact-checking, and to Martha Cipolla for a fantastic copyedit, which also included some fact-checking. Now I understand why they always say this: any remaining mistakes are my own.

In addition to those quoted in the book, I interviewed a number of other scientists, experts, and advocates, all of whom have my gratitude. Special thanks go to Andrew Orrell.

For feedback on the book, for friendship, for other support, or all of the above, my thanks to Mona Gable, Joel Sappel, Mark Engler, Genevieve Scott, Amy Powell, Herschel Farbman, Amy DePaul, Cascade Sorte,

Sabine Kunrath, Deepa Ranganathan, Soren Tjernell, John Mangin, Michelle Tupko, Josh Rolnick, Seth Rosenthal, and, especially, Rosalie Metro.

I am fortunate to be a friend of UC Irvine's literary journalism program and affiliated with the Humanities Center. Thank you to Amanda Swain, Angelica Enriquez, Ekua Arhin, Judy Wu, Barry Siegel, Hector Tobar, and Jeff Wasserstrom for various kinds of support and opportunities, as well as friendship. Finally, Erika Hayasaki has shaped me and my work more than she knows. Her extraordinary generosity and inexhaustible appetite for discussing craft, and our frequent sessions of "journalism therapy," have been a gift.

My deep gratitude goes to Dan, for being the best brother ever; to Jane and Paul, for countless reasons; and to my parents, for planting in me the seeds of the questions animating this book, and in so many ways making it possible for me to write it.

The questions about the future of our planet acquired extra weight for me when I became a parent. How can I begin to thank Eliza, to whom this book is dedicated? It's been the honor of my life to watch you become who you are. Lucky for me, who you are is someone with excellent literary judgment. Thank you for your crucial contribution to improving this book, and to enriching my life.

Nicholas, your singular mind, exacting standards, and total candor make you a great editor and interlocutor. Other qualities, notably your genius for goofiness, make you a great person with whom to spend a life. After twenty-odd years, I'm not bored yet.

NOTES

Introduction

1. Lois Beckett, "'It's Too Hot': Los Angeles Melts under its Worst Heatwave of the Year," *The Guardian*, September 3, 2022, https://www.theguardian.com/us-news/2022/sep/02/los-angeles-extreme-heatwave-emergency.

Haley Smith and Brittny Mejia, "Extreme Heat Waves are Making Firefighters Sick, adding new dangers to job," *Los Angeles Times*, September 2, 2022, https://www.latimes.com/california/story/2022-09-02/heat-injuries-are-a-growing-threat-to-wildfire-crews.

2. Video of the Senate floor session can be found in the California State Senate Media Archive: https://www.senate.ca.gov/media-archive?title=&start_date=2022-08-31&end_date=2022-08-31.

3. Lewis Strauss, chairman of the Atomic Energy Commission, said this in a speech in 1954. The full speech can be found here: https://www.nrc.gov/docs/ML1613/ML16131A120.pdf.

4. "The Diablo Canyon Blockade," *It's About Times*, Abalone Alliance newspaper, October/November 1981, reproduced in *Found SF*, https://www.foundsf.org/index.php?title=The_Diablo_Canyon_Blockade_1981.

"N-protesters Gird for New Assault," *San Francisco Sunday Examiner and Chronicle*, September 20, 1981, p. A1.

5. "Most U.S. Nuclear Power Plants Were built between 1970 and 1990," U.S. Energy Information Administration, April 27, 2017, https://www.eia.gov/todayinenergy/detail.php?id=30972.

6. There were a few antecedents, such as Bruno Comby, who published a book called *Environmentalists for Nuclear Energy* in the 1990s, but they did not gain much traction at the time.

7. Michael Weisskopf, "Scientist Says Greenhouse Effect Is Setting In," *Washington Post*, June 23, 1988, https://www.washingtonpost.com/archive/politics/1988/06/24/scientist-says-greenhouse-effect-is-setting-in/3844f00f-42f4-420f-8811-62de6c989d8f/.

8. James Hansen, Kerry Emanuel, Ken Caldeira, and Tom Wigley, "Nuclear Power Paves the Only Viable Path Forward on Climate Change," *The Guardian*, December 3, 2015, https://www.theguardian.com/environment/2015/dec/03/nuclear-power-paves-the-only-viable-path-forward-on-climate-change.

9. There have been several documentaries featuring pro-nuclear advocates, including *Pandora's Promise, Atomic Hope*, and *Nuclear Now*. These are very much worth watching,

but I see them more as a manifestation of the pro-nuclear movement—their producers include pro-nuclear donors and advocates–than as strictly journalistic coverage of it.

10. See Derek Thompson, "A Simple Plan to Solve All of America's Problems," *The Atlantic*, January 12, 2022, https://www.theatlantic.com/ideas/archive/2022/01/scarcity-crisis-college-housing-health-care/621221/.

11. The anti-nuclear advocate was Chelsi Sparti, a graduate student at UC Berkeley.

Chapter One

1. In 1977, Scott earned a teaching credential from Cal Poly in industrial education and construction tecnhologies.

Information about Scott Lathrop comes from author interviews with him and documents he provided.

2. Peter H. King, "Brown Vows He'll Try to Stop Diablo," *San Francisco Examiner*, July 1, 1979, A1.

3. The ownership arrangements of the Diablo Canyon Lands are rather complicated. In general in this book, I will refer to the land as owned by PG&E, although parts are owned by subsidiaries.

4. Brian F. Codding and Terry L. Jones, *Foragers on America's Western Edge: The Archaeology of California's Pecho Coast*, University of Utah Press, 2019, 8.

PG&E submission to the Atomic Energy Commission, 1973, and "Diablo Canyon Lands," Diablo Canyon Decommissioning Engagement Panel, https://diablocanyonpanel.org/decom-topics/diablo-canyon-lands/.

5. Per correspondence with Dr. John Johnson, they were referred to as Obispeño for more than a century, until the 1980s.

6. Codding and Jones, *Foragers on America's Western Edge*, 32.

7. Author interview with Dr. John Johnson.

8. John R. Johnson, "Descendants of Native Ranchérias in the Diablo Lands Vicinity: A Northern Chumash Ethnohistorical Study" (report commissioned by PG&E), 7, and private materials from Scott Lathrop.

9. Roberta Greenwood, "9000 Years of Prehistory at Diablo Canyon, San Luis Obispo County, California," San Luis Obispo County Archaeological Society, 1972.

10. "Indians Protest Diablo Plant," *New York Times* (UPI), Oct. 6, 1981, section A, page 29.

Chapter Two

1. My descriptions of the march are based on video and other materials provided by Heather Hoff. I also retraced the steps of the march myself to observe the location.

2. There had been rallies in 2010 by a small group of supporters of a threatened plant in Vermont. This book focuses on California, but another book could be written about both anti-nuclear and pro-nuclear activism in the Northeast.

3. Sarah Spath's name is now Sarah Woolf.

4. James E. Lovelock and Lynn Margulis, "Atmospheric Homeostasis by and for the Biosphere: The Gaia Hypothesis," *Tellus* 26, no. 1–2 (1974): 2–10.

5. James Lovelock, "Nuclear Power Is the Only Green Solution," *The Independent*,

May 24, 2004, https://www.independent.co.uk/voices/commentators/james-lovelock-nuclear-power-is-the-only-green-solution-564446.html.

6. "Life Cycle Greenhouse Gas Emissions from Electricity Generation: Update," National Renewable Energy Lab, September 2021, https://www.nrel.gov/docs/fy21osti/80580.pdf.

7. "Fort Calhoun Becomes Fifth U.S. Nuclear Plant to Retire in Past Five Years," U.S. Energy Information Administration, October 31, 2016, https://www.eia.gov/todayinenergy/detail.php?id=28572.

8. Biographical information about Shellenberger's youth comes from author interview with Shellenberger. Information in the rest of the paragraph comes from author interviews with Shellenberger, Ted Nordhaus, and Eric Meyer.

9. The range of opinion about Nordhaus and Shellenberger in the environmental community can be seen especially clearly in coverage by *Grist*, the environmentally themed outlet, which gave them extensive coverage starting with the publication of "Death of Environmentalism." The coverage was often quite critical, but the publication also gave them a platform and took them seriously, and some of the coverage was relatively sympathetic. As for the range of opinion on Shellenberger within the pro-nuclear community, this is based on author interviews with numerous members of the movement. See chapters eight and sixteen for more details.

10. For an excellent exploration of the YIMBY movement, see Conor Dougherty, *Golden Gates: Fighting for Housing in America* (New York: Penguin Press, 2020).

11. At the time, Heather Hoff's name was Heather Matteson, her married name. She changed her name back to Hoff after a divorce. For simplicity's sake, I refer to her as Hoff throughout the book.

Chapter Three

For this chapter, I have benefited from the insights of those who have written extensively about the history of the Sierra Club, especially Thomas Wellock, Tom Turner, and Susan Schrepner.

1. Bill McKibben, ed., *American Earth: Environmental Writing Since Thoreau* (New York: Literary Classics of the Unites States, 2008), 98.

2. Tom Turner, *David Brower: The Making of the Environmental Movement* (Oakland, CA: University of California Press, 2015), 9–44.

3. William E. Siri, "Reflections on the Sierra Club, the Environment, and Mountaineering, 1950s–1970s," Oral History Center, The Bancroft Library, University of California, Berkeley, 2.

4. Dougherty, *Golden Gates*, 66.

5. There is some dispute over how low-carbon hydroelectric power actually is. It varies by facility. In terms of reliability, as with other energy sources, it has started to be affected by climate change, which sometimes diminishes hydro resources.

6. Turner, *David Brower*, 123–4.

7. "Einstein's Letter," The Manhattan Project: An Interactive History, U.S. Department of Energy, https://www.osti.gov/opennet/manhattan-project-history/Events/1939-1942

/einstein_letter.htm. The letter was drafted by Einstein and fellow scientist Leo Szilard, but signed only by Einstein.

8. Patricia Rife, *Lise Meitner and the Dawn of the Nuclear Age* (Boston: Birkhauser, 1999), 256.

9. Turner, *David Brower*, 188.

10. J. Samuel Walker and Thomas R. Wellock, "A Short History of Nuclear Regulation, 1946–2009," U.S. Nuclear Regulatory Commission.

11. "The New Explorers: Atoms for Peace," TV documentary, 1996. Thank you to Yoon Chang for providing me with the DVD of this episode. The early economics of nuclear power were very complex, and outside the scope of this book. For an excellent analysis of the topic, see James Jasper, *Nuclear Politics: Energy and the State in the United States, Sweden, and France* (Princeton, New Jersey: Princeton University Press, 1990).

12. Historical materials provided by PG&E.

13. Quoted in Thomas Raymond Wellock, *Critical Masses: Opposition to Nuclear Power in California, 1958–1978* (Madison: University of Wisconsin Press, 1988), 17.

14. Sierra Club archives, Bancroft Library, carton 79, folder 10.

15. Wellock, *Critical Masses*, 58. Although most of the Sierra Club board was not against nuclear power per se at this time, Pesonen was.

16. Wellock, *Critical Masses*, 80.

17. Wellock, *Critical Masses*, 81.

18. Barry Commoner, *The Closing Circle: Nature, Man, and Technology* (New York: Alfred A. Knopf, 1972), 50–51.

19. Quoted in Frank Graham, *Since Silent Spring* (Houghton Mifflin, 1970), 13. I don't believe this passage by Carson was ever published; Graham writes only, "She put some of her anxieties down on paper[.]" I am indebted to Spencer Weart for bringing my attention to it.

20. These details come from contemporaneous press coverage.

21. Turner, *David Brower*, 105.

22. Rachel Carson, *Silent Spring* (New York: Houghton Mifflin Company, 2002, originally published 1962), 6.

23. Susan R. Schrepfer, "Diablo Canyon and the Transformation of the Sierra Club, 1965–1985," *California History Magazine*, Summer 1992, Volume LXXI, No. 2.

24. Siri, "Reflections on the Sierra Club, the Environment, and Mountaineering, 1950s–1970s," 108.

25. Paul Ehrlich, *The Population Bomb* (Ballantine Books, 1968), 1.

26. Turner, 9.

27. Turner, *David Brower*, 80. Michael Shellenberger has also written about Brower's reasons for opposing nuclear.

28. Gladwin Hill, "Brower Quitting Sierra Club Post: Conservation Group Leader Plans New Organization," *New York Times*, retrieved from https://www.proquest.com/historical-newspapers/brower-quitting-sierra-club-post/docview/118622303/se-2.

29. Turner, *David Brower*, 149.

30. Schrepfer, "Diablo Canyon and the Transformation of the Sierra Club, 1965–1985."

31. National Review Mission Statement, William F. Buckley Jr., https://www.nationalreview.com/1955/11/our-mission-statement-william-f-buckley-jr/.

32. In recent years, there has been extensive debate about the racial attitudes of John Muir and other early Sierra Club members as well. See Michael Brune, "Pulling Down Our Monuments," Sierra Club website, July 22, 2020, https://www.sierraclub.org/michael-brune/2020/07/john-muir-early-history-sierra-club.

Chapter Four

1. Correspondence with PG&E.

2. "Putting the Axe to the Scram Myth," U.S. NRC Blog, May 17, 2011, https://www.nrc.gov/reading-rm/basic-ref/students/history-101/putting-axe-to-scram-myth.html.

3. "Report of the President's Commission on the Accident at Three Mile Island: The Need for Change: The Legacy of TMI," October 1979, Washington, D.C.

4. "Capacity Factor of Nuclear Power Plants in the United States from 1975 to 2023," Statista, June 28, 2024, https://www.statista.com/statistics/191201/capacity-factor-of-nuclear-power-plants-in-the-us-since-1975/.

5. "What Is Generation Capacity?" Office of Nuclear Energy, May 1, 2020, https://www.energy.gov/ne/articles/what-generation-capacity. Thanks to Steve Nesbit for drawing my attention to this point in an interview.

6. David R. Schiel, John R. Steinbeck, and Michael S. Foster, "Ten Years of Induced Ocean Warming Causes Comprehensive Changes in Marine Benthic Communities," *Ecology* 85, no. 7 (2004): 1833-1839.

7. "Cancer Statistics," National Cancer Institute, https://www.cancer.gov/about-cancer/understanding/statistics.

8. David B. Richardson, Klervi Leuraud, Dominique Laurier, Michael Gillies, Richard Haylock, Kaitlin Kelly-Reif, Stephen Bertke, et al., "Cancer Mortality After Low Dose Exposure to Ionising Radiation in Workers in France, the United Kingdom, and the United States (INWORKS): Cohort Study," *Bmj* 382 (2023); David B. Richardson, Elisabeth Cardis, Robert D. Daniels, Michael Gillies, Jacqueline A. O'Hagan, Ghassan B. Hamra, Richard Haylock, et al., "Risk of Cancer from Occupational Exposure to Ionising Radiation: Retrospective Cohort Study of Workers in France, the United Kingdom, and the United States (INWORKS)," *Bmj* 351 (2015).

Chapter Five

1. Her full name is Lucy Jane Swanson, but she has always gone by Jane.

2. National Archives, "Vietnam War U.S. Military Fatal Casualty Statistics," https://www.archives.gov/research/military/vietnam-war/casualty-statistics.

3. Susan Sontag, "The Imagination of Disaster," *Essays of the 1960s & 1970s*, Library of America, 2013, 207–8.

4. John W. Gofman and Arthur R. Tamplin, *Poisoned Power: The Case Against Nuclear Power Plants*, Emmaus, PA: Rodale Press, 1971, 21.

5. Gofman and Tamplin, *Poisoned Power*, 23.

6. Gofman and Tamplin, *Poisoned Power*, 64.

7. Gofman and Tamplin, *Poisoned Power*, 25.

8. Gofman and Tamplin, *Poisoned Power*, 74.

9. Gofman and Tamplin, *Poisoned Power*, 123.

10. "Fallout and Disarmament: A Debate between Linus Pauling and Edward Teller." *Daedalus* 87, no. 2 (1958): 147–63. http://www.jstor.org/stable/20026443.

11. "Battle Cry Building Across U.S," *Newsday* (1940–); June 12, 1978; ProQuest Historical Newspapers: *Newsday*, 4.

12. "Founding Document: 1968 MIT Faculty Statement," Union of Concerned Scientists, https://www.ucsusa.org/about/history/founding-document-1968-mit-faculty-statement.

13. Steve Esmesdina, "Benefit Concert: Jackson Browne Sings for a Cause," *Los Angeles Times*, January 29, 1979, p. SD4.

14. Steve Morse, "Singers Fight Nukes: No Issue Since Vietnam Has United Them So Much," *Boston Globe*, June 3, 1979; p. A10.

15. Michael Unger, "Battle Cry Building Across U.S.," *Newsday*, June 12, 1978, p. 4.

16. Dorothy Nelkin, "Nuclear Power as a Feminist Issue," Environment: Science and Policy for Sustainable Development, 23, no. 1 (1981): 14–39, DOI: 10.1080/00139157.1981.9940928

17. Ibid.

18. Susan Koen and Nina Swaim, *Aint No Where We Can Run: Handbook for Women on the Nuclear Mentality* (Trumansburg, NY: The Crossing Press, 1980), 2.

19. Dorothy Nelkin, "Nuclear Power as a Feminist Issue," *Environment: Science and Policy for Sustainable Development* 23, no. 1 (1981): 14–39.

20. Robert A. Stallings, "Evacuation at Three Mile Island," *International Journal of Mass Emergencies and Disasters*, 1984.

21. "Report of the President's Commission on the Accident at Three Mile Island: The Need for Change: The Legacy of TMI," October 1979, Washington, D.C.

22. I have assembled this account from contemporaneous newspaper coverage, as well as the Abalone Alliance's own publication, *It's About Times*. See "The Diablo Canyon Blockade 1981: 'I was there . . . ,'" by Ward Young and Mark Evanoff, with help from many others, *It's About Times*, October–November 1981, https://www.foundsf.org/index.php?title=The_Diablo_Canyon_Blockade_1981.

23. Judith Cummings, "Coast A-Plant Construction Error Tied to Missing Guide to Blueprint," *New York Times*, October 2, 1981, p. A14.

24. "Our History Timeline," Mothers for Peace, https://mothersforpeace.org/about-us/our-history/

25. Wellock, *Critical Masses*, 29.

26. Wellock, *Critical Masses*, 13.

27. "State Restrictions on New Nuclear Power Facility Construction," National Conference on State Legislatures, https://www.ncsl.org/environment-and-natural-resources/states-restrictions-on-new-nuclear-power-facility-construction.

Chapter Six

1. "A Special Series on the Alleged 'Death of Environmentalism,'" *Grist*, January 14, 2005, https://grist.org/politics/doe-intro/.

2. Patrick Allitt, *A Climate of Crisis: America in the Age of Environmentalism* (New York: Penguin Press, 2014), 73.

3. Bill McKibben, *The End of Nature* (New York: Random House, 2006, originally published in 1989), 7.

4. Technically, he won the Sveriges Riksbank Prize in Economic Sciences in Memory of Alfred Nobel.

5. Nordhaus told me that Shellenberger always saw himself as a world historical figure. In an email, I told Shellenberger that Nordhaus had described him this way. He did not respond.

6. The account of Bracken Hendricks's contribution comes from author interview with Peter Teague. Hendricks could not be reached for comment.

7. It was also known as the Apollo Alliance. The name evolved over time.

8. In addition to Nordhaus, Shellenberger, and Teague, I interviewed Carl Pope, Dan Becker, and Ralph Cavanagh. The latter were members of the mainstream environmental community. The 10-point plan did include some language on regulation, but it was not emphasized. See http://www.cce-mt.org/archives/HGS%20Plant/HGS%20FEIS/Studies/files/Apollo%20Jobs%20Report.pdf.

9. Quoted in Michael Shellenberger and Ted Nordhaus, "The Death of Environmentalism: Global Warming Politics for a Post-Environmental World," 28.

10. Shellenberger and Nordhaus, "Death of Environmentalism," 12.

11. Shellenberger and Nordhaus, "Death of Environmentalism," 9.

12. As a friend who speaks Chinese explained to me, "One part of the compound that makes up crisis does mean 'danger' [but] the other part does not mean 'opportunity' but rather is one part of a different compound that means that. . . . It's just one of those not quite true things that took on a life of its own."

13. Carl Pope, "An In-Depth Response to 'The Death of Environmentalism,'" *Grist*, January 14, 2005, https://grist.org/article/pope-reprint/ (accessed December 28, 2022).

14. Bill McKibben, "Bill McKibben Sends Dispatches from a Conference on Winning the Climate-Change Fight," *Grist*, January 26, 2005, https://grist.org/politics/mckibben3/.

15. Felicity Barringer, "Paper Sets Off a Debate on Environmentalism's Future," *New York Times*, February 6, 2005, p. 1.18.

16. David Roberts, "Why I've Avoided Commenting on Nisbet's 'Climate Shift' Report," *Grist*, April 27, 2011, https://grist.org/climate-change/2011-04-26-why-ive-avoided-commenting-on-nisbets-climate-shift-report/.

17. https://www.youtube.com/watch?v=qwIW6LmEDAU.

18. Ted Nordhaus and Michael Shellenberger, *Break Through: From the Death of Environmentalism to the Politics of Possibility* (New York: Houghton Mifflin Company, 2007), 98.

Chapter Seven

1. This apparently changed at some point. As of this writing, the Bill and Melinda Gates Foundation website has a section called "Water, Sanitation & Hygiene," which notes, "Inadequate sanitation and hygiene are estimated to have caused more than half a million deaths from diarrhea alone in 2016."

2. Another concept that sounds almost magical is nuclear fusion—which would mimic the behavior of stars, producing energy by fusing atoms together rather than splitting them apart. Fusion is beyond the scope of this book.

3. "History of INL," Idaho National Laboatory, https://inl.gov/history/.

4. Charles E. Till and Yoon Il Chang, *Plentiful Energy: The Story of the Integral Fast Reactor*, 2011, 58.

5. Till and Chang, *Plentiful Energy*, 1. I have removed italics from the original quote.

6. "The New Explorers: Atoms for Peace," TV documentary. Thanks to Yoon Chang for sending me the DVD. Dr. Chang also explained how the reactor worked and provided additional details.

7. Ibid.

8. I have seen different estimates; this is what Yoon Chang told me was a ballpark figure.

9. Mitchell Locin, "Senators Save Argonne Project," *Chicago Tribune*, July 1, 1994, p. 1.

10. Mitchell Locin, "Congress Pulls Plug on 10-year Argonne Project," *Chicago Tribune*, August 5, 1994, p. SW1.

11. Tom Blees, *Prescription for the Planet: The Painless Remedy for Our Energy and Environmental Crises*, BookSurge Publishing, 2008, 115.

12. Amory B. Lovins, "Energy Strategy: The Road Not Taken," *Foreign Affairs* 55, no. 1 (October 1976): 65-96.

13. James Hansen, "Trip Report," https://www.columbia.edu/~jeh1/mailings/2008/20080804_TripReport.pdf.

14. James Hansen, *Storms of My Grandchildren: The Truth about the Coming Climate Catastrophe and Our Last Chance to Save Humanity* (New York: Bloomsbury, 2011), 202–203.

15. Hansen, *Storms of My Grandchildren*, 203.

16. Blees, *Prescription for the Planet*, 117.

17. https://www.columbia.edu/~jeh1/mailings/2008/20081229_DearMichelleAndBarack.pdf. A bit more detail about Hansen's recommendations: he urged the phasing out of coal-fired plants that do not capture and store CO_2, and in addition to a tax, he urged a dividend—that is, using the tax revenue to return money to households.

Chapter Eight

1. "Great East Japan Earthquake and Tsunami," United Nations Environment Programme, https://www.unep.org/topics/disasters-and-conflicts/country-presence/japan/great-east-japan-earthquake-and-tsunami

2. "Out of Control: Merkel Gambles Credibility with Nuclear U-Turn," Spiegel staff, translated by Christopher Sultan, *Der Spiegel*, March 21, 2011, https://www.spiegel.de/international/germany/out-of-control-merkel-gambles-credibility-with-nuclear-u-turn-a-752163.html, accessed March 2, 2023.

3. Nordhaus and Shellenberger, *Break Through*, 271.

4. Fred Turner, *From Counterculture to Cyberculture: Stewart Brand, the Whole Earth Network, and the Rise of Digital Utopianism* (Chicago: University of Chicago Press, 2008), 71.

5. Stewart Brand, *Whole Earth Discipline: An Ecopragmatist's Manifesto* (Viking Adult, 2009), 21.

6. "5 Fast Facts about Spent Nuclear Fuel," Department of Energy, October 3, 2022, https://www.energy.gov/ne/articles/5-fast-facts-about-spent-nuclear-fuel

7. Brand, *Whole Earth Discipline*, 80.

8. Quoted in Brand, *Whole Earth Discipline*, 88.

9. E. Broughton, "The Bhopal Disaster and its Aftermath: A Review," *Environ Health*, 2005 May 10; 4(1):6. doi: 10.1186/1476-069X-4-6.

10. "Frequently Asked Chernobyl Questions," International Atomic Energy Agency, https://www.iaea.org/newscenter/focus/chernobyl/faqs.

11. Brand, *Whole Earth Discipline*, 81.

12. Quoted in Brand, *Whole Earth Discipline*, 86.

13. Brand, *Whole Earth Discipline*, 20.

14. Brand, *Whole Earth Discipline*, 1.

15. Amory Lovins, "Stewart Brand's Nuclear Enthusiasm Falls Short on Facts and Logic," *Grist*, October 14, 2009, https://grist.org/article/2009-10-13-stewart-brands-nuclear-enthusiasm-falls-short-on-facts-and-logic/.

16. Brand confirmed this in an email to me.

17. Denyse O'Leary, "Science and Society: Here a Tic, There a Tic, Everywhere a Heretic," *Uncommon Descent* (blog), October 1, 2008, https://uncommondescent.com/science/science-and-society-here-a-tic-there-a-tic-everywhere-a-heretic/.

18. John Markoff, *Whole Earth: The Many Lives of Stewart Brand* (New York: Penguin Press, 2022), 354. Brand confirmed in an email to me that it was a very bad night.

19. George Monbiot, "Why Fukushima Made Me Stop Worrying and Love Nuclear Power," *The Guardian*, March 21, 2011, https://www.theguardian.com/commentisfree/2011/mar/21/pro-nuclear-japan-fukushima, accessed November 28, 2023.

20. Jessica Lovering, Ted Nordhaus, and Michael Shellenberger, "Out of the Nuclear Closet: Why it's Time for Environmentalists to Stop Worrying and Love the Atom," *Foreign Policy*, September 7, 2012, https://foreignpolicy.com/2012/09/07/out-of-the-nuclear-closet/ (accessed November 28, 2023).

21. David Roberts, "Why I've Avoided Commenting on Nisbet's 'Climate Shift' Report," *Grist*, April 27, 2011, https://grist.org/climate-change/2011-04-26-why-ive-avoided-commenting-on-nisbets-climate-shift-report/.

22. Author interview with Alex Trembath.

23. In an email, I informed Shellenberger of what his former colleagues had said, to give him the opportunity to respond. He did not reply to the email.

24. The text of the toast is from an email from Ted Nordhaus to author.

25. Peter Teague had joined the Breakthrough Institute at this point and was involved in negotiations with Shellenberger. After Shellenberger left, he was executive director "for about five minutes," he told me, before he left and Nordhaus took over. In author interviews, Teague also corroborated Nordhaus's characterizations of the challenges of working with Shellenberger at this time.

Chapter Nine

1. S. David Freeman, *The Green Cowboy: An Energetic Life* (Authorhouse: 2016), 145. For this chapter, in addition to Freeman's autobiography, I drew on interviews with Ralph Cavanagh, William Manheim, Tom Dalzell, Damon Moglen, and others.

2. John Schwartz, "S. David Freeman, 94, Tireless Advocate for Clean Energy, Dies," *New York Times*, May 17, 2020.

3. Freeman, *The Green Cowboy*, 27.

4. Freeman, *The Green Cowboy*, 86.

5. Tom Zeller Jr., "U.S. Nuclear Plants Have Same Risks, and Backups, as Japan Counterparts," *New York Times*, March 13, 2011, A10.

6. Freeman, *The Green Cowboy*, 147. Regarding the causes of the closure, there was also a theory about financial impropriety on the part of CPUC chairman Michael Peevey, who was also a friend of Freeman's.

7. I could not track down the original source of the quote, so it may be apocryphal. Here is one source that invokes it: Alexander Cockburn and Jeffrey St. Clair, "The Ring of Eternal Fire," *The Ecologist*, November 11, 2013, https://theecologist.org/2013/nov/11/ring-eternal-fire.

8. "Report on the Analysis of the Shoreline Fault Zone, Central Coastal California: Report to the U.S. Nuclear Regulatory Commission," PG&E, January 2011, https://www.nrc.gov/docs/ML1101/ML110140425.pdf.

9. V. John White and Associates, "A Cost Effective and Reliable Zero Carbon Replacement Strategy for Diablo Canyon Power Plant," https://foe.org/wp-content/uploads/2017/legacy/PlanBfinal.pdf.

10. Amory B. Lovins, "Energy Strategy: The Road Not Taken," *Foreign Affairs* 55, no. 1 (October 1976), 88.

11. "U.S. Solar Market Sets New Record, Installing 7.3 GW of Solar PV in 2015," Solar Energy Industries Association, February 19, 2016, https://www.seia.org/news/us-solar-market-sets-new-record-installing-73-gw-solar-pv-2015#:~:text=The%20U.S.%20installed%207%2C286%20MW,five%20years%20earlier%20in%202010.

12. "2015 Renewable Energy Data Book," U.S. Department of Energy, https://www.nrel.gov/docs/fy17osti/66591.pdf.

13. Morgan Lee, "Natural Gas Replacing Nuclear Power in SD," *San Diego Union-Tribune*, February 19, 2015, https://www.sandiegouniontribune.com/sdut-natural-gas-replaces-nuclear-2015feb19-story.html.

14. Katherine Blunt, *California Burning: The Fall of Pacific Gas and Electric—and What It Means for America's Power Grid* (Portfolio, 2022), 98.

15. Dana Hull, "California Hits Renewable Milestone: 1 Gigawatt of Solar Power Installed to Date," *The Mercury News*, November 8, 2011, https://www.mercurynews.com/2011/11/08/california-hits-renewable-energy-milestone-1-gigawatt-of-solar-power-installed-to-date/#:~:text=One%20gigawatt%20is%20roughly%20the,than%20France%2C%20China%20and%20Belgium.

16. "What is Generation Capacity?" Department of Energy, May 1, 2020, https://www.energy.gov/ne/articles/what-generation-capacity.

17. Interview with Bill Manheim of PG&E.

18. The State Lands Commission did not technically need to approve the deal itself, but rather to renew a lease that was necessary to carry out the plan.

19. Interview with Tom Dalzell.

20. S. David Freeman, "Diablo Canyon Agreement Should Put an End to Debate About Nuclear Power," *The Sacramento Bee,* June 25, 2016, https://www.sacbee.com/opinion/california-forum/article85614127.html.

Chapter Ten

1. Karn Vohra, Alina Vodonos, Joel Schwartz, Eloise A. Marais, Melissa P. Sulprizio, and Loretta J. Mickley, "Global Mortality from Outdoor Fine Particle Pollution Generated by Fossil Fuel Combustion: Results from GEOS-Chem," *Environmental Research* 195 (2021): 110754. (A more recent study puts the number at "only" around 5 million.)

2. A good overview of what is known about the death toll of Chernobyl can be found at https://ourworldindata.org/what-was-the-death-toll-from-chernobyl-and-fukushima.

3. "Diablo Canyon Lives," *Decouple* podcast, September 2, 2022.

4. "Q&A," Mothers for Nuclear, https://www.mothersfornuclear.org/faq.

5. For example, see "Nuclear Power: Low-Carbon Electricity with Serious Economic and Safety Issues," Union of Concerned Scientists, https://www.ucsusa.org/energy/nuclear-power.

6. My account is based on video of the meeting and the transcript. The transcript can be found here: https://www.slc.ca.gov/Meeting_Transcripts/2016_Documents/06-28-2016_Transcripts.pdf.

Chapter Eleven

1. I would later learn, however, that a trained opera singer also worked on nuclear energy as a technician at Idaho National Lab.

2. This is what Eric remembers; I was unable to confirm with Shellenberger.

3. Jackie Kempfer is now Jackie Siebens.

4. See, for example, "Americans Divided on Nuclear Energy," Gallup, May 20, 2022, https://news.gallup.com/poll/392831/americans-divided-nuclear-energy.aspx.

5. "Radiation Effects and Sources," United Nations Environment Programme, 2016.

6. "Early Estimate of Motor Vehicle Traffic Fatalities in 2023," National Highway Traffic Safety Administration, April 2024, https://crashstats.nhtsa.dot.gov/Api/Public/ViewPublication/813561.

7. "Safety Record of U.S. Air Carriers," Airlines for America, December 24, 2022.

8. For example, Edwin Lyman of the Union of Concerned Scientists told me he believes the death toll is higher, and the Richardson papers cited above suggest that working at a nuclear plant can impose additional cancer risks. There is also evidence that pilots and flight attendants have higher rates of certain kinds of cancers. See "Aircrew Safety & Health," CDC, https://www.cdc.gov/niosh/aviation/prevention/index.html.

9. https://www.mothersfornuclear.org/our-thoughts/2018/3/11/fukushima-perspective fromareactoroperatoron3/11-7yearslater.

10. Ibid.

11. "Radiation: Health Consequences of the Fukushima Nuclear Accident," World Health

Organization, March 10, 2016, https://www.who.int/news-room/questions-and-answers/item/health-consequences-of-fukushima-nuclear-accident.

12. "Fukushima Daiichi Accident FAQ," World Nuclear Association, https://world-nuclear.org/focus/fukushima-daiichi-accident/fukushima-daiichi-accident-faq.aspx.

13. Hiroko Tibuchi, "Japan Races to Build New Coal-Burning Power Plants, Despite the Risks," *New York Times*, February 3, 2020, https://www.nytimes.com/2020/02/03/climate/japan-coal-fukushima.html; see also Pushker A. Kharecha, and Makiko Sato, "Implications of energy and CO_2 emission changes in Japan and Germany after the Fukushima accident," *Energy Policy* 132 (2019): 647–653.

14. Philip Thomas, "Evacuating a Nuclear Disaster Area Is (Usually) a Waste of Time and Money, Says Study," *The Conversation*, November 20, 2017, https://theconversation.com/evacuating-a-nuclear-disaster-areas-is-usually-a-waste-of-time-and-money-says-study-87697.

15. Arifumi Hasegawa, Koichi Tanigawa, Akira Ohtsuru, Hirooki Yabe, Masaharu Maeda, Jun Shigemura, Tetsuya Ohira, et al., "Health Effects of Radiation and Other Health Problems in the Aftermath of Nuclear Accidents, with an Emphasis on Fukushima," *The Lancet* 386, no. 9992 (2015): 479–488.

16. For a powerful look at the psychological effects of radiation, see Kai Erikson, *A New Species of Trouble: The Human Experience of Modern Disasters* (W.W. Norton & Company, 1994).

17. Adam Higginbotham, *Midnight in Chernobyl* (Simon & Schuster, 2019), 335.

18. Leah Stokes, "While the Planet Overheats, Ohio's Coal Industry Gets a Bailout," *The Guardian*, July 28, 2019, https://www.theguardian.com/commentisfree/2019/jul/28/planet-overheats-ohios-coal-industry-gets-a-bailout.

19. Hansen and Kharecha, "Nuclear Power Saves Lives," *Nature* 497, 539 (2013), https://doi.org/10.1038/497539e.

20. Pushker A. Kharecha, and Makiko Sato, "Implications of Energy and CO_2 Emission Changes in Japan and Germany After the Fukushima Accident," *Energy Policy* 132 (2019): 647-653.

21. https://twitter.com/shellenberger/status/1153742232203120640.

22. Michael Shellenberger, "If Renewables Are So Great for the Environment, Why Do They Keep Destroying It?" *Forbes*, May 17, 2018, https://www.forbes.com/sites/michaelshellenberger/2018/05/17/if-renewables-are-so-great-for-the-environment-why-do-they-keep-destroying-it/?sh=31913c3c3a1c.

23. See "The Wrong Way to Save Ohio's Zero-Carbon Nuclear Plants," Josh Freed and Ryan Fitpatrick, Third Way, June 7, 2019, https://www.thirdway.org/blog/the-wrong-way-to-save-ohios-zero-carbon-nuclear-plants.

24. Laurel Wamsley, "Ohio Speaker Arrested in Connection with $60 Million Bribery Scheme," July 21, 2020, NPR, https://www.npr.org/2020/07/21/893493224/ohio-house-speaker-arrested-in-connection-to-60-million-bribery-scheme.

Chapter Twelve

1. "Panel Formation," Diablo Canyon Decommissioning Engagement Panel, https://diablocanyonpanel.org/about-us/formation-and-history/.

2. Interview with Dr. John Johnson.

3. Interview with Scott Lathrop, Mona Tucker, and Wendy Lucas. They said that there was more to the name than that, but it was special knowledge to the tribe. It has also been translated as "the people of San Luis Obispo."

4. David Treuer, "Return the National Parks to the Tribes," *The Atlantic*, May 2021 issue, https://www.theatlantic.com/magazine/archive/2021/05/return-the-national-parks-to-the-tribes/618395/.

5. https://www.gov.ca.gov/2019/06/18/governor-newsom-issues-apology-to-native-americans-for-states-historical-wrongdoings-establishes-truth-and-healing-council/.

6. Jill Cowan, "'It's Called Genocide': Newsom Apologizes to the State's Native Americans," *New York Times*, June 19, 2019, https://www.nytimes.com/2019/06/19/us/newsom-native-american-apology.html.

7. https://landback.org.

8. Much of the material about the Navajo experience comes from Judy Pasternak, *Yellow Dirt* (New York: Free Press, 2010).

9. Doug Brugge and Rob Goble, "The History of Uranium Mining and the Navajo People." *American Journal of Public Health* 92, no. 9 (2002): 1410-1419.

10. "Indigenous Anti-Nuclear Summit Declaration," Indigenous Environmental Network, https://www.ienearth.org/indigenous-anti-nuclear-summit-declaration/#:~:text=September%205%2D8%2C%201996,deadly%20effects%20on%20our%20communities.

11. "A Strategic Vision," Diablo Canyon Decommissioning Engagement Panel, December 2018 (updated most recently April 2023).

Chapter Thirteen

1. The task force's report can be found here: https://www.samuellawrencefoundation.org/_files/ugd/6a4539_599ab6493fbd4036bbac6e73ad1123d7.pdf.

2. National Research Council. 1957. *Disposal of Radioactive Waste on Land; Report.* Washington, D.C.: The National Academies Press. https://doi.org/10.17226/18527.

3. Leonard Ross, "How 'Atoms for Peace' Became Bombs for Sale," *New York Times*, December 5, 1976, p. 240. I also interviewed Edwin Lyman and Tom Isaacs about proliferation history and risk. It's important to note that to produce a weapon from spent fuel, a reprocessing facility would be needed. Also, according to Isaacs, it is easier to produce a weapon from a research reactor than from a power plant. He says that no commercial nuclear power plant has been used to produce a weapon.

4. "Long-Lived Fission Products," radioactivity.eu.com, https://radioactivity.eu.com/articles/nuclearenergy/long_lived_fission_products#; and Madison Hilly, "Nuclear Waste Is Misunderstood," *New York Times*, April 28, 2023, https://www.nytimes.com/2023/04/28/opinion/climate-change-nuclear-waste.html. This piece is by a pro-nuclear advocate, but I asked a nuclear waste expert, Tom Isaacs, if the estimate seemed reasonable, and he said yes.

5. Even Edwin Lyman of Union of Concerned Scientists, who is known for raising questions about nuclear safety, told me he is unaware of any reported cases of harm from spent nuclear fuel.

6. Hilly, "Nuclear Waste Is Misunderstood."

7. https://twitter.com/MadiHilly/status/1671491294831493120.

8. Lisa Mascaro, "Obama Administration: 'We're done with Yucca,'" *Las Vegas Sun*, January 29, 2010, https://lasvegassun.com/news/2010/jan/29/obama-administration-were-done-yucca-mountain/.

9. "Yucca Mountain Repository," U.S. Department of Energy, https://web.archive.org/web/20080822205714/http://www.ocrwm.doe.gov/ym_repository/index.shtml#4.

10. "Blue Ribbon Commission on America's Nuclear Future: Report to the Secretary of Energy," January 2012, https://www.energy.gov/ne/articles/blue-ribbon-commission-americas-nuclear-future-report-secretary-energy.

11. There are two sections of the ISFSI with slightly different designs; the older one is farther from the shore and designed for inundation by 50 feet of water.

12. "Radiation from Air Travel," Centers for Disease Control and Prevention, https://www.cdc.gov/radiation-health/data-research/facts-stats/air-travel.html?CDC_AAref_Val=https://www.cdc.gov/nceh/radiation/air_travel.html.

Chapter Fourteen

1. Sadly, in May of 2024, Mother's Tavern closed.

2. Peter Miller, "Diablo Canyon Accord Approved by California Legislature," August 20, 2018, https://www.nrdc.org/bio/peter-miller/diablo-canyon-accord-approved-california-legislature.

3. Thomas L. Neff, "A Grand Uranium Bargain," *New York Times*, October 24, 1991, A25.

4. In addition to reading published articles about the program, I interviewed Thomas Neff.

5. "More Megatons to Megawatts," Bulletin of the Atomic Scientists, Dawn Stover, February 21, 2014, https://thebulletin.org/2014/02/more-megatons-to-megawatts/; Khalil Ryan, "Megatons to Megawatts: An Explainer," Good Energy Collective, December 20, 2023, https://www.goodenergycollective.org/policy/megatons-to-megawatts-an-explainer.

6. Oddly to me, after my initial conversations with Gene, he became more of a climate skeptic, and has even written in support of natural gas.

7. "Amory Lovins: Expanding Nuclear Power Makes Climate Change Worse," Democracy Now!, July 16, 2008, https://www.democracynow.org/2008/7/16/amory_lovins_expanding_nuclear_power_makes.

8. At some point since the '60s, the *San Luis Obispo County Telegram-Tribune* had dropped the "County" and the "Telegram."

9. https://www.sanluisobispo.com/opinion/letters-to-the-editor/article136547208.html.

10. I have not been able to track down the original quote, but I have found it attributed to Snyder in several sources: *Courting the Wild Twin*, by Martin Shaw, p. 91 and *Confessions of a Recovering Environmentalist and Other Essays*, by Paul Kingsnorth, p. 105.

11. "State Options to Keep Nuclear in the Energy Mix," National Conference of State Legislatures, January 2017.

12. Ivan Penn, "PG&E Ordered to Pay $3.5 Million for Causing Deadly Fire,"

New York Times, April 6, 2021, https://www.nytimes.com/2020/06/18/business/energy-environment/pge-camp-fire-sentenced.html?searchResultPosition=2.

13. Carl Wurtz told me that CGNP asked Shellenberger to chip in for a lobbyist. Shellenberger told me that he had hired a lobbyist. When I later emailed Shellenberger to confirm that he had split the cost with CGNP, he did not reply.

Chapter Fifteen

1. Video of the event can be viewed here: https://www.youtube.com/watch?v=C122qJH_Hwg.

2. "AR5 Climate Change 2014: Mitigation of Climate Change," IPCC, 2014, https://www.ipcc.ch/report/ar5/wg3/.

3. Sam Brinton, "The Advanced Nuclear Industry," Third Way, June 15, 2015, https://www.thirdway.org/report/the-advanced-nuclear-industry.

4. Jackie Toth and Jackie Kempfer, "How Advanced Nuclear Got on the Map," Third Way, April 8, 2021, https://www.thirdway.org/memo/how-advanced-nuclear-got-on-the-map.

5. Ted Nordhaus and Kenton de Kirby, "Nuclear Cognition: Public Attitudes, Elite Opinion, and the Next Generation of Nuclear Energy Communications," Breakthrough Institute, 2021.

6. Toth and Kempfer, "How Advanced Nuclear Got on the Map."

7. https://foe.org/wp-content/uploads/2019/01/Progressive-Climate-Leg-Sign-On-Letter-2.pdf

8. Jacqueline Toth, "Ocasio-Cortez: Green New Deal 'Leaves the Door Open' on Nuclear," Morning Consult Pro, May 6, 2019, https://pro.morningconsult.com/articles/ocasio-cortez-green-new-deal-leaves-door-open-nuclear.

9. Peter Behr, "Biden, Once a Critic, May Boost Nuclear Power," E&E News, December 3, 2020, https://www.eenews.net/articles/biden-once-a-critic-may-boost-nuclear-power/; see platform here: https://joebiden.com/wp-content/uploads/2020/08/UNITY-TASK-FORCE-RECOMMENDATIONS.pdf.

10. "The Nuclear Power Dilemma," Union of Concerned Scientist, October 9, 2018, https://www.ucsusa.org/resources/nuclear-power-dilemma.

11. 2020 Democratic Party Platform.

Chapter Sixteen

1. Michael Shellenberger, "On Behalf of Environmentalists, I Apologize for the Climate Scare," Environmental Progress website, June 29, 2020, https://environmentalprogress.org/big-news/2020/6/29/on-behalf-of-environmentalists-i-apologize-for-the-climate-scare.

2. Michael Shellenberger, "Why Earth Overshoot Day and the Ecological Footprint Are Pseudoscientific Nonsense," *Forbes*, July 29, 2019, https://www.forbes.com/sites/michaelshellenberger/2019/07/29/why-earth-overshoot-day-and-the-ecological-footprint-are-pseudoscientific-nonsense/?sh=6cf821237e76.

3. Michael Shellenberger, "California Has Always Had Fires, Environmental Alarmism Makes Them Worse Than Necessary," *Forbes*, September 10, 2020, https://www.forbes.com

/sites/michaelshellenberger/2020/09/10/why-environmental-alarmism-makes-forest-fires-worse/?sh=2f5650693712.

4. Michael Shellenberger, "Don't Deny Plastics to Poor Nations," *Forbes*, August 31, 2020, https://www.forbes.com/sites/michaelshellenberger/2020/08/31/dont-deny-plastics-to-poor-nations/?sh=5be8ff007531.

5. Michael Shellenberger, "If They Are So Alarmed by Climate Change, Why Are They Opposed to Solving It?", *Forbes*, February 17, 2020, https://www.forbes.com/sites/michaelshellenberger/2020/02/17/if-they-are-so-alarmed-by-climate-change-why-are-they-so-opposed-to-solving-it/?sh=475ee3ce6b75.

6. The groups founded by Madison Hilly and Paris Ortiz-Wines would likely not exist without Shellenberger because he converted them to the pro-nuclear cause. He did not convert Mark Nelson, but he helped him set up Radiant Energy Group. Hilly told me that she established her organization independently, declining an offer of financial support from Shellenberger.

7. Michael Shellenberger, "Why Biden Can Unite America with Nuclear Power—Or Divide It with Renewables," *Forbes*, https://www.forbes.com/sites/michaelshellenberger/2020/11/09/why-biden-can-unite-america-with-nuclear-power—-or-divide-it-with-renewables/.

8. Jessica Lovering called him loud, and Ted Nordhaus called him a narcissist. The other quotes were not for attribution.

9. I mentioned to Shellenberger in an email that I believed he was recording me for the entirety of the conversation. He did not reply.

10. "End-of-Life Solar Panels: Regulations and Management," U.S. EPA, https://www.epa.gov/hw/end-life-solar-panels-regulations-and-management.

11. Akela Lacy, "South Carolina Spent $9 Billion to Dig a Hole in the Ground and Then Fill It Back in," *The Intercept*, February 6, 2019, https://theintercept.com/2019/02/06/south-caroline-green-new-deal-south-carolina-nuclear-energy/. The reactors referenced in this paragraph are known as AP1000. They are designed to be simpler and more compact than previous generations of reactors, and to incorporate passive safety features. However, they are still large, light-water reactors, which is why I refer to them as the "old style."

12. Michael Shellenberger, *Apocalypse Never: Why Environmental Alarmism Hurts Us All* (Harper, 2020), 5–6.

13. Alex Trembath, "Alternatives to Climate Alarmism," *National Review*, August 10, 2020 issue, https://www.nationalreview.com/magazine/2020/08/10/alternatives-to-climate-alarmism/.

14. "An Ecomodernist Manifesto," April 2015, http://www.ecomodernism.org.

15. Hannah Ritchie and Pablo Rosado, "Natural Disasters," Our World in Data, https://ourworldindata.org/natural-disasters.

16. "Article by Michael Shellenberger Mixes Accurate and Inaccurate Claims in Support of a Misleading and Overly Simplistic Argumentation About Climate Change," Climate Feedback, July 6, 2020, https://climatefeedback.org/evaluation/article-by-michael-shellenberger-mixes-accurate-and-inaccurate-claims-in-support-of-a-misleading-and-overly-simplistic-argumentation-about-climate-change/.

17. Author interview with Kerry Emmanuel.

Chapter Seventeen

1. Alex P. Meshik, "The Workings of an Ancient Nuclear Reactor," *Scientific American*, November 1, 2005, https://www.scientificamerican.com/article/the-workings-of-an-ancien/.

2. George A. Cowan, "A Natural Fission Reactor," *Scientific American*, Vol. 235, No. 1 (July 1976), pp. 36–47.

3. "Western Reliance on Russian Fuel: A Dangerous Game," Third Way, September 20, 2023, https://www.thirdway.org/memo/western-reliance-on-russian-fuel-a-dangerous-game.

4. The IFR and EBR-2 are sometimes referred to interchangeably, which can be confusing. The IFR was a concept, a prototype of which was built in EBR-2. Oklo was awarded spent fuel from EBR-2, which may have included fuel from the IFR as well as other fuel.

5. NRC press release: https://www.nrc.gov/reading-rm/doc-collections/news/2022/22-002.pdf.

6. These analysts were Ed Lyman, Tom Isaacs, and Ross Matzkin-Bridger.

7. Ralph Nader and John Abbotts, *The Menace of Atomic Energy* (W.W. Norton & Company, 1977), 257.

Chapter Eighteen

1. Isabelle's relationship with Joe Gebbia is public knowledge and was initially mentioned to me by Heather Hoff. Isabelle referred to her friendships with Elon Musk and Grimes in a podcast episode: "Behind the Diablo Canyon Victory feat. Isabelle Boemeke," *Decouple* podcast, September 7, 2022, https://www.decouple.media/p/behind-the-diablo-canyon-victory-6e4.

2. CGNP has a website, Gene was active on Facebook and LinkedIn, and more recently started writing a Substack.

3. The letter can be found here: https://drive.google.com/file/d/1wpKHHQD9IVHDQOQPLMFYIHPt5x0kLm0_/view.

4. Isabelle Boemeke, "Nuclear Power Is Our Best Hope to Ditch Fossil Fuels," TED Talk, April 2022, https://www.ted.com/talks/isabelle_boemeke_nuclear_power_is_our_best_hope_to_ditch_fossil_fuels?language=en&subtitle=en.

5. Justin Aborn, Ejeong Baik, et al., "An Assessment of the Diablo Canyon Nuclear Plant for Zero-Carbon Electricity, Desalination, and Hydrogen Production," November 2021.

6. Steven Chu and Ernest Moniz, "California Needs to Keep Diablo Canyon Nuclear Plant Open to Meet Its Climate Goals," *Los Angeles Times*, November 21, 2021, https://www.latimes.com/opinion/story/2021-11-21/diablo-canyon-nuclear-plant-climate-change-zero-emissions.

7. "CEC Determines Diablo Canyon Power Plant Needed to Support Grid Reliability," California Energy Commission, February 28, 2023, https://www.energy.ca.gov/news/2023-02/cec-determines-diablo-canyon-power-plant-needed-support-grid-reliability.

8. Sammy Roth, "What caused California's rolling blackouts? Climate change and poor planning," *Los Angeles Times*, October 6, 2020, https://www.latimes.com/environ

ment/story/2020-10-06/california-rolling-blackouts-climate-change-poor-planning#:~:text=The%20rotating%20power%20outages%20didn,as%202%C2%BD%20hours%20on%20Aug.

9. Sammy Roth, "California Promised to Close Its Last Nuclear Plant. Now Newsom Is Reconsidering," *Los Angeles Times*, April 29, 2022, https://www.latimes.com/environment/story/2022-04-29/california-promised-to-close-its-last-nuclear-plant-now-newsom-is-reconsidering.

10. Dianne Feinstein, "Why I Changed My Mind About California's Diablo Canyon Nuclear Plant," *Sacramento Bee*, June 15, 2022, https://www.sacbee.com/opinion/op-ed/article262499997.html.

Chapter Nineteen

1. Hans Rosling, "The Magic Washing Machine," TED Talk, December 2010, https://www.ted.com/talks/hans_rosling_the_magic_washing_machine.

2. I spoke with the intern, Jordan, and he denied that he had been interviewed as part of this investigation.

3. "About," Good Energy Collective, https://www.goodenergycollective.org/about-us.

4. David Roberts, "Nuclear Power Has Been Top-Down and Hierarchical. These Women Want to Change That," Vox, July 21, 2020, https://www.vox.com/energy-and-environment/2020/7/21/21328053/climate-change-nuclear-power-environmental-justice-energy-collective.

5. https://twitter.com/amywestervelt/status/1442993777770074119.

6. As far as I can tell, this term originated with the Breakthrough Institute.

7. Seth Feaster, "U.S. on Track to Close Half of Coal Capacity by 2026," Institute for Energy Economics and Financial Analysis, April 3, 2023, https://ieefa.org/resources/us-track-close-half-coal-capacity-2026.

Chapter Twenty

1. "As Heat Wave Grips Western U.S., Governor Newsom Takes Action to Increase Energy Supplies and Reduce Demand," https://www.gov.ca.gov/2022/08/31/as-heat-wave-grips-western-u-s-governor-newsom-takes-action-to-increase-energy-supplies-and-reduce-demand/.

2. "Diablo Canyon Lives," *Decouple* podcast, September 2, 2022.

3. I later learned that cockroaches were a symbol of the Cold War and nuclear apocalypse, often represented as the sole survivors. A fascinating episode of the *Radio Atlantic* podcast explores this symbolism and how, just as the Cold War was ending, an effective way to get rid of cockroaches was invented. This is interesting context to keep in mind when contemplating how nuclear attitudes have shifted. "How We Turned the Tide in the Roach Wars," *Radio Atlantic* podcast, November 30, 2023.

Epilogue

1. "Growing Share of Americans Favor More Nuclear Power," Pew Research Center, August 18, 2023.

2. "State Restrictions on New Nuclear Power Facility Construction," National

Conference on State Legislatures, https://www.ncsl.org/environment-and-natural-resources/states-restrictions-on-new-nuclear-power-facility-construction.

3. See, for example, Intergovernmental Panel on Climate Change, "Climate Change 2022: Mitigation of Climate Change," 2022, https://www.ipcc.ch/report/ar6/wg3/downloads/report/IPCC_AR6_WGIII_FullReport.pdf; and International Energy Agency, "Nuclear Power and Secure Energy Transitions," June 2022, https://www.iea.org/reports/nuclear-power-and-secure-energy-transitions.

4. "Chernobyl: The True Scale of the Accident," World Health Organization, September 5, 2005, https://www.who.int/news/item/05-09-2005-chernobyl-the-true-scale-of-the-accident.

5. Yeager M, Machiela MJ, Kothiyal P, Dean M, Bodelon C, Suman S, Wang M, Mirabello L, Nelson CW, Zhou W, Palmer C, Ballew B, Colli LM, Freedman ND, Dagnall C, Hutchinson A, Vij V, Maruvka Y, Hatch M, Illienko I, Belayev Y, Nakamura N, Chumak V, Bakhanova E, Belyi D, Kryuchkov V, Golovanov I, Gudzenko N, Cahoon EK, Albert P, Drozdovitch V, Little MP, Mabuchi K, Stewart C, Getz G, Bazyka D, Berrington de Gonzalez A, Chanock SJ. "Lack of Transgenerational Effects of Ionizing Radiation Exposure from the Chernobyl Accident," *Science*, 2021 May 14;372(6543):725-729. doi: 10.1126/science.abg2365.

6. Gofman and Tamplin, *Poisoned Power*, 18.

7. Power Reactor Information System, International Atomic Energy Agency, "Nuclear Share of Electricity Generation in 2023," https://pris.iaea.org/pris/worldstatistics/nuclearshareofelectricitygeneration.aspx.

8. Turner, *David Brower*, 161.

9. Bryan Walsh, "How the Sierra Club Took Millions from the Gas Industry—and Why They Stopped," *Time*, February 2, 2012, https://science.time.com/2012/02/02/exclusive-how-the-sierra-club-took-millions-from-the-natural-gas-industry-and-why-they-stopped/.

10. ProPublica Nonprofit Explorer, "Sierra Club," https://projects.propublica.org/nonprofits/organizations/941153307.

11. ProPublica Nonprofit Explorer, "Natural Resources Defense Council Inc," https://projects.propublica.org/nonprofits/organizations/132654926.

12. Frederick Soddy, *The Interpretation of Radium and the Structure of the Atom* (New York: Putnam, 1922), 8. At the time of the lecture series, he had not yet won the Nobel Prize; he won in 1921.

13. Charlie Hoffs, "Mining Raw Materials for Solar Panels: Problems and Solutions," October 19, 2022, https://blog.ucsusa.org/charlie-hoffs/mining-raw-materials-for-solar-panels-problems-and-solutions/.

14. Angela Symons, "Watch Greta Thunberg Being Carried Away by Police During Anti-Wind Farm Protest in Norway," Euronews, https://www.euronews.com/green/2023/03/01greta-thunberg-says-norways-wind-farms-are-an-unacceptable-violation-of-human-rights.

15. Lyric Aquino, "Biden Administration Pauses Copper Mining Project on Oak Flat, a Sacred Apache Site," *Grist*, May 24, 2023, https://grist.org/indigenous/biden-copper-mine-arizona-put-on-pause-oak-flat.

16. Ezra Klein, "The Economic Mistake Democrats Are Finally Confronting," *New York Times*, September 19, 2021, https://www.nytimes.com/2021/09/19/opinion/supply-side-progressivism.html; Ezra Klein, "What America Needs Is a Liberalism That Builds," *New York Times*, May 29, 2022, https://www.nytimes.com/2022/05/29/opinion/biden-liberalism-infrastructure-building.html. I also interviewed Bhaskar Sunkara.

17. Josh Lappen, "The Climate Coalition Is Threatening to Split Apart," *Heatmap*, June 22, 2023, https://heatmap.news/politics/the-climate-coalition-is-threatening-to-split-apart?utm_id=106015&sfmc_id=5002687.